19th International Heat Pipe Conference and 13th International Heat Pipe Symposium (IHPC 2018)

IOP Conference Series: Materials Science and Engineering
Volume 1139

Pisa, Italy
10 - 14 June 2018

Editors:

Sauro Filippeschi
Marco Marengo

ISBN: 978-1-7138-3414-4
ISSN: 1757-8981

Printed from e-media with permission by:

Curran Associates, Inc.
57 Morehouse Lane
Red Hook, NY 12571

Some format issues inherent in the e-media version may also appear in this print version.

Institute of Physics
Dirac House, Temple Back
Bristol BS1 6BE UK

Phone: 44 1 17 929 7481
Fax: 44 1 17 920 0979

techtracking@iop.org

Additional copies of this publication are available from:

Curran Associates, Inc.
57 Morehouse Lane
Red Hook, NY 12571 USA
Phone: 845-758-0400
Fax: 845-758-2633
Email: curran@proceedings.com
Web: www.proceedings.com

TABLE OF CONTENTS

THERMOSYPHONS

IOP Conf. Series: Materials Science and Engineering 1139 (2021) 011001 doi:10.1088/1757-899X/1139/1/011001

Selected papers of the Joint 19th International Heat Pipe Conference and 13th International Heat Pipe Symposium

Sauro Filippeschi[1*] and Marco Marengo[2]

1 University of Pisa, Department of Energy, Systems, TErritory and Constructions Engineering (DESTEC), Pisa, Italy
2 University of Brighton, Advanced Engineering Centre, University of Brighton, Brighton, United Kingdom

The Joint 19th IHPC (International Heat Pipe Conference) and the 13th IHPS (International Heat Pipe Symposium) was held at the University of Pisa, June 10th-14th, 2018. The joint conference focuses on topics related to heat pipes, thermosyphons and other two-phase thermal devices. The conference is the most important and attended scientific and networking event in the field of two-phase heat transfer devices worldwide. The 1st International Heat Pipe Symposium was held at Tokyo, Japan, in 1985, and it has been held in staggered years from the IHPC until 2016, when the first joint event has been organized in Jeju Island, South Korea. The conference held at Pisa, has been the second joint event of this important conference and welcomed a large number of scientists and technologists from all over the world.

Two-phase thermal devices perform excellent heat transfer rates with very low thermal differences, they are passive and they can operate on ground and in space environment. They can be applied to the micro heat transfer applications as well as the large size devices. All these technical peculiarities make these devices largely applied to different engineering fields: from the electronic cooling to the thermal control of sensors, battery and other miscellaneous devices from the energy saving in industrial and residential applications to the renewable energy exploitation. Despite their large use, the two phase devices he driven physical mechanisms are not totally understood and new kind of device different from the original heat pipes [1] or thermosyphons have been developed and studied. This conference therefore, encourages the exchange of ideas between scientists working on such different engineering topics as evaporation, boiling, condensation, two-phase fluid dynamic in capillary pipes and porous structures and lastly, working on natural fluid circulation. Furthermore, the conference inspires the collaboration among scientists and technologists in order to: provide design tools for heat pipe builders, solve technological issues as the chemical compatibility, the aging and safety, and lastly, optimise the performances of heat pipes and thermosyphon for specific

* Corresponding author: sauro.filippeschi@unipi.it

applications. A total number of 195 persons attended the conference coming from 22 different countries. The permanent International Heat Pipe Committee preliminarily selected 162 abstracts from a total number of 184 works and 141 full papers were submitted to the conference. The most of them (95 papers) have been orally presented, the rest of papers have been discussed in two posters sections.

A selection of the papers dealing with microgravity and space applications have been published in a topical collection of the Journal Microgravity Science and Technology [2], in a special issue of the ASME Journal of Heat Transfer [3] in the Journal Thermal Science and Engineering Progress. All the papers submitted to the conference have been preliminarily reviewed by the International Heat Pipe Committee. After a further review process, in agreement with IOP Conference Serie Policy, a selection of 15 papers is now published. They deal with standard Heat Pipes (5 papers), Pulsating Heat Pipes (2 papers), Loop Heat Pipes and Capillary Pumped Loops (2 papers), Thermosyphons (3 papers), and, finally, 3 papers deal with a generic two-phase thermal device.

We wish to thank the International Heat Pipe Committee and the Local Organizing Committee for their great support and advices.

Guest Editors
Prof. Sauro Filippeschi
Prof. Marco Marengo

References

[1] Grover G. M., Evaporation and Condensation Device, U.S. Patent N°3,229,759, 1963
[2] Khandekar S., Savino R., Heat pipe systems for thermal management in space, Topical Collection Microgravity Science and Technology, Springer ISSN: 0938-0108 (Print) 1875-0494 (Online) , 2019
[3] Bozzoli F., Groll M., Rossetto L., Special Issue: Joint 19th IHPC and 13th IHPS, *J. Heat Transfer*. September 2019, 141(9): 090301. doi: https://doi.org/10.1115/1.4044493

International Heat Pipe Committee

Prof. W. Qu, Beijing, China
Prof. P. Stephan, Darmstadt, Germany
Mr. W. Supper, Noordwijk, The Netherlands
Prof. P. Terdtoon, Chiang Mai, Thailand
Prof. L. L. Vasiliev, Minsk, Belarus

Local Organizing Committee

Chair
Prof. Marco Marengo, University of Brighton, UK
Co-chair
Prof. Sauro Filippeschi, University of Pisa, Italy
Members
Dr. Mauro Mameli, University of Pisa, Italy
Prof. Fabio Fantozzi, University of Pisa, Italy
Prof. Paolo Di Marco, University of Pisa, Italy
Prof. Fabio Bozzoli, University of Parma, Italy
Dr. Lucio Araneo, Politecnic of Milan, Italy
Prof. Lucia Rossetto, University of Padova, Italy
Prof. Raffaele Savino, Univerisity Federico II Naples, Italy
Dr. Andrea Mariani, Enea, Casaccia, Italy
Dr. Andrea Merlone, Inrim C.N.R., Turin, Italy
Dr. Filomena Iorizzo, Argotec, Turin, Italy
Dr. Marco Molina, Leonardo S.p.A, Italy
Dr. Lorenzo Caporale, AAVID Thermalloy, Italy
Honorary members
Dr. Carlo Bassani, ISPRA, Leggiuno, Italy
Prof. Enrico Maria Latrofa, University of Pisa, Italy

IOP Conf. Series: Materials Science and Engineering 1139 (2021) 011002 doi:10.1088/1757-899X/1139/1/011002

Peer review declaration

All papers published in this volume of *IOP Conference Series: Materials Science and Engineering* have been peer reviewed through processes administered by the Editors. Reviews were conducted by expert referees to the professional and scientific standards expected of a proceedings journal published by IOP Publishing.

- **Type of peer review:** Single-anonymous: authors' identities are known to the reviewers, reviewers' identities are hidden from authors

- **Describe criteria used by Reviewers when accepting/declining papers. Was there the opportunity to resubmit articles after revisions?**

  The criteria used by the reviewers to accept or decline a paper are listed before and can be divided into two areas: **ethical and technical**.
  For the **ethical** area the reviewer had checked if this paper doesn't contain: biased considerations in regard to race, gender, religious belief, ethnic origin, citizenship or political philosophy of the authors or offensive terminology or words might be perceived to be offensive to others.
  For the **technical area** the referee had given a clear statement of recommendation according to the following scheme:

  - **Scientific merit** (scientific rigour, accuracy of the scientific text, correctness of the scientific contents
  - **Clarity of expression** (clarity in driving ideas communication, paper reading fluency, concepts organization)
  - **Discussion of the context of the work**: (complete context discussion)
  - **Referencing** (suitability of the references)
  - **Quality Criteria (**originality, motivation, scientific interest of the main topic, parts of manuscript repetition, suitability of the length)
  - **Presentation Criteria** (title adequacy, Abstract lengths and completeness, Diagrams, figures, tables and captions clarity, Text and mathematics adequacy and clarity)
  - **Conclusions** (adequacy and completeness of the conclusions)

  After a first review process the paper could be resubmitted until the reviewers or the guest editors accepted it.

- **Conference submission management system:**

No special system was used. The papers were submitted to the conference secretariat and then distributed to reviewers by the editors. The reviewers sent their reviews to the editors. The editors manage the whole editing process.

- **Number of submissions received:** 17 papers

- **Number of submissions sent for review:** 17 papers

- **Number of submissions accepted:** 15 papers

- **Acceptance Rate (Number of Submissions Accepted / Number of Submissions Received X 100):** 88,23%

- **Average number of reviews per paper:** 2 reviewers for papers

- **Total number of reviewers involved:** 30 reviewers

  Any additional info on review process (ie plagiarism check system): The plagiarism check has been done directly by the reviewer supervised by the editors.

- **Contact person for queries:**

Prof. Sauro Filippeschi
DESTEC – University of Pisa
sauro.filippeschi@unipi.it

IOP Conf. Series: Materials Science and Engineering 1139 (2021) 012001 doi:10.1088/1757-899X/1139/1/012001

Study of diffusion bonded flat plate closed loop pulsating heat pipes with alternating porous media

Luis Betancur[1*], Arthur Facin[1], Priscila Gonçalves[1], Kleber Paiva[2], Marcia Mantelli[1,] and Gustavo Nuernberg[1]

[1]Department of Mechanical Engineering, Federal University of Santa Catarina, Florianopolis, 88040-900, Brazil
[2]Department of Mobility Engineering, Federal University of Santa Catarina, Joinville, 89219-600, Brazil

Abstract. This work aims to discuss the application of diffusion bonding process in the fabrication of flat plate pulsating heat pipes and to analyze the thermal performance effect of porous media structures applied to alternating channels in the evaporator section and compare it with a smooth conventional flat plate pulsating heat pipe. For that, two different copper flat plate pulsating heat pipes with five turns and 1.5 mm of inner diameter channels were fabricated, using diffusion bonding. Samples were made with two different diffusion process to evaluate the ability of the resulting device to keep the vacuum, avoiding the use of filler material during the welding process, which could result in contamination and, therefore, undesirable corrosion on both internal and external heat pipe surfaces. The results allowed to assess the quality of the bonded interface of two different configurations of the diffusion process. Microscopic results obtained on controllable pressure furnace and non-pressure control were analyzed, where the diffusion process with controlled pressure presented best results of continuity, alignment, and tightness. A comparison of the heat transfer characteristics between the two devices was performed under the same experimental conditions; for both devices, the thermal resistance in horizontal position showed to be like heat transfer of pure conductive copper plate. However, pulsating heat pipe containing smooth channels showed the lower thermal resistance in vertical position and filling ratio of 50%.

1. Introduction

Pulsating heat pipes (PHP) are highly efficient heat exchanger devices, consisting of an evacuated small diameter closed tube bended in multiple turns, in which a certain amount of working fluid is inserted. PHPs are composed of two main regions: evaporator, where heat is inserted, and condenser, where heat is removed from, as schematically presented in Figure 1. An adiabatic section, a thermally insulated region, located between the evaporator and the condenser, may have a variable dimension, or even not exist. PHPs operate in closed two-phase chaotic cycles. Confined saturated liquid, when heated, forms slugs of liquid and plugs of vapor as shown in Figure 1. Due to the confinement, PHPs use latent heat of vaporization to generate vapor, expanding the bubbles and pushing the liquid slugs to the condenser region. The high-speed cycles are responsible for the efficiency of the working fluid latent and sensible heat transportation.

* corresponding author: luisbetancur@labtucal.ufsc.br

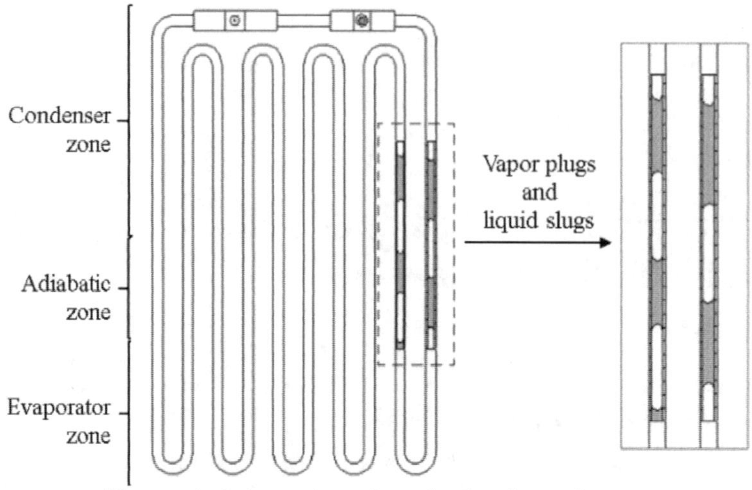

Figure 1. Schematics of a pulsating heat pipe.

When heat power is provided to the evaporator zone, vapor is generated and, due to pressure perturbations, transported along the pipeline to be rejected in the condenser zone. Two simultaneous phenomena occur in a PHP: bubble-growing (and pressure increase) at the evaporator and bubble-collapsing at the condenser, inducing the pressure decrease in this region. This growing-collapsing process produces the fluctuation of the liquid slugs, trapped between the vapor plugs. Heat is transferred by two means: liquid-vapor and vapor-liquid phase change and sensible heat, due to the displacement of liquid slugs [1] [2]. PHPs are a promising solution for heat flux management, since they are wickless, easy to manufacture and present fewer operating limitations.

Some researchers studied the thermal performance of closed loop PHP by introducing valves in the circuit or by increasing the number of turns. Although efficient, check valves increase the manufacturing complexity of the device [3]. Moreover, the increase of the number of turns can improve the thermal performance of PHP; however, it can represent a geometry limitation in actual applications. According to [4], the fabrication simplicity of a PHP contrasts with the complex thermo-hydro-dynamic operation principles that drive its operation and which include the effect of the surface tension dominance over the two phase bubble liquid slug formation inside the tube.

PHP can be classified between meandered closed loop and flat plate PHP. The second is a device that contains channels on a planar surface, welded against another channeled or flat plate. In this case, the heat is also transferred by conduction through the plate where the channels are made, besides the pulsating effect of the working fluid. Among the advantages, the design of flat plate PHP is almost unlimited, as practically any channel geometry can be achieved, and the channels surface can be easily modified before joining both plates. The alteration of chemical composition and/or properties of the surface, such as roughness or porosity, directly affects friction and wettability of the working fluid with the channel. Different forms of capillary unbalancing have been reported by different authors, who concluded that this type of device is less sensitive to the inclination angle. Also, the device's thermal resistance decreased with the increase of the heat input and the vertical arrangement showed a lower thermal resistance when compared to horizontal positions [3] [5] [6].

Diffusion bonding in solid state is a technique that produces coalescence of surfaces in intimate contact, by the application of a pressing force during a thermal cycle under vacuum. The main mechanism of the joining process is the interdiffusion of atoms across the surfaces, after the plastic deformation of surface asperities [7]. Such technique relies on the combination of three main variables: temperature, pressing load and holding time, which vary according to the material pair, its surface finishing, and the expected service conditions. Unlike conventional processes, the diffusion bonding preserves the base metal microstructure and properties at the interface, since it can be performed without filler materials, localized thermal gradient nor melting. However, the surfaces to be joined require more preparation than for conventional welding processes; whole parts are subjected to longer thermal cycles and equipment costs are usually high [8]. Diffusion bonding was used in this work to bond commercial copper plates in the fabrication of two different flat plate closed loop PHPs. A conventional one and another with asymmetrical sintered porous media onto alternating channels at the evaporator. The first one was fabricated using a commercial diffusion bonding furnace and the second one using a conventional resistive tubular furnace. This work compares the thermal performance of two flat plate PHPs produced with and without porous media and evaluates the resulting microstructure of the channels with surface modification and of the joint regions for the different processing routes.

2. Fabrication process

Two flat plate PHP of 5 U-turns formed by two machined copper plates were fabricated using different experimental apparatus for the diffusion bonding, detailed in sequence.

Semi cylindrical grooves with 1.5mm of diameter was milled by CNC in two flat plates (see Figure 2a) and united against each other, to form circular cross section channels. To remove surface contaminants, the parts were cleaned by immersion in a $10_{vol\%}$ aqueous solution of H_2SO_4. In sequence, the planar surfaces to be bonded were grinded with silicon carbide paper, with fine grit size (1500), to obtain the smoothness required to assure uniform contact. Eight holes were drilled to host removable steel pins of 3mm of diameter, used to avoid misalignment of plates during the diffusion bonding process. The available contact area for diffusion bonding in each device was of 3,862mm^2.

2.1. Conventional flat plate pulsating heat pipe

After surface preparation, the diffusion bonding of the conventional flat plate PHP was performed in a high vacuum heat treatment furnace with an integrated pressing unit, PVA TePla MOV 653HP. The copper plates were positioned in the furnace and submitted to a thermal cycle, as detailed in Figure 2b and c. A pressure of 8MPa was applied to the device during the diffusion bonding thermal cycle, under vacuum of at least 5x10^{-5}mbar and followed by slow cooling inside the furnace (Figure 3).

a) Machined copper plates. b) Stacked plates c) Device inside the furnace.

Figure 2. Plates preparation and assembly before the diffusion bonding cycle

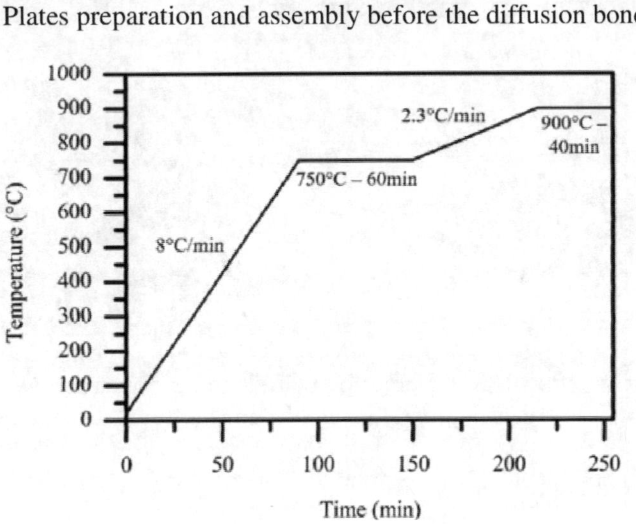

Figure 3. Diffusion bonding thermal cycle.

2.2. Flat plate pulsating heat pipe containing porous media in alternating channels

The second flat plate PHP also consists of two machined plates with the same geometry as the other one, but with modified (wicked) channels by sintering particulate copper at the evaporator zone, as indicated in Figure 4a and b. The step consisting of sintering loose copper powder over the surface of the channels was performed before the diffusion bonding of the copper plates. No particle deformation occurs during the loose powder sintering process, a pressureless shaping technique of powder metallurgy, which favor the manufacture of highly porous materials or layers [9]. A commercial copper powder (PAC-Metalpó) was sieved and the particles retained in sieve with 400μm mesh size were selected and poured in alternating channels of the evaporator, forming a layer with 40mm of length and around 0.25mm of thickness.

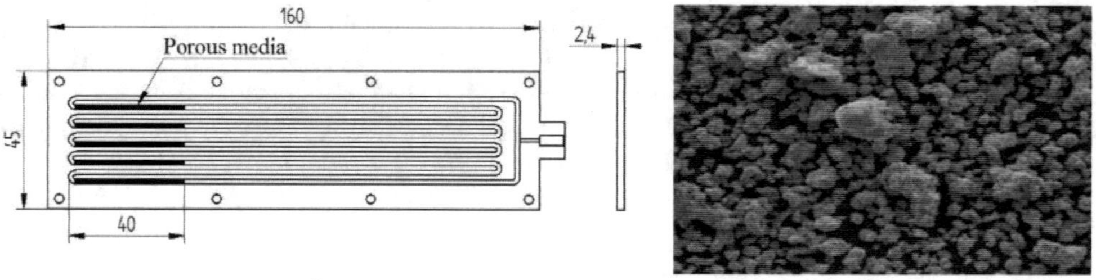

a) b)

Figure 4. Porous media distribution; a) Overall dimensions and porous media distribution at evaporator zone; b) Typical Cu particle morphology and size distribution.

The concave format was obtained using a graphite core of 1mm of dimeter, laid up over the poured powder, as presented in Figure 5a. The sintering of powder was performed in a resistive furnace (Figure 5b), at 900ºC during one hour and under controlled atmosphere of 95%Ar and 5%H_2. Heating was programmed to increase in a rate of 10°C/minute. Despite all the benefits of using a commercial furnace for diffusion bonding, the elevated energy consumption, operational costs, and time-consuming cycles must be considered. As an alternative, the diffusion bonding was performed using a resistive furnace with the help of a pressure applying matrix, a device consisting of two stainless steel thick plates with passing screws near its edges to provide the needed pressure for intimate contact of the surfaces [10].

The flat plates with modified channels were positioned between the boards hold by fourteen M10 screws, which applied an initial pressure of 10bar to the copper plates by hydraulic press. The whole system was placed inside the furnace Figure 5b and heated up to 900ºC maintained for one hour, using a heating rate of 10ºC per minute, and cooled naturally inside the furnace.

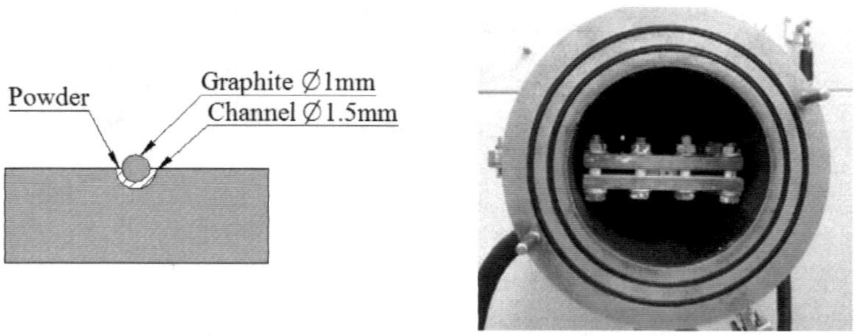

a) b)

Figure 5. Powder distribution; a) Graphite core over the powder bed.
b) plates-screws assembly inside the conventional resistive furnace.

After the fabrication of both devices, the vacuum leakage was tested up to 10^{-7}mbar.l/s to evaluate if the fabrication process used was able to produce sealed bonded surfaces.

3. Experimental setup

For charging the working fluid, the devices were first vacuumed by a vacuum pump (Edwards RV8) up to 1x10^{-6} mbar. An Edwards Spectron 5000 Leak Detector equipment checked the sealing of the device.

The working fluid was first degassed by continuous vacuuming cycles, by utilizing a volumetric pump. The temperature distribution of the flat plate PHP was measured by eight type T thermocouples (±0.9°C of maximum uncertainty), distributed along the device, according to Figure 6a. A DAQ-NI SCXI-1000 data acquisition system was used to record experiment data. In the condenser region, heat was removed from the PHP by an aluminium heat sink, as shown in Figure 6b. This heat sink has two channels within which water flows from a thermal bath (Lauda® Proline RP1845). This equipment can keep the temperature at constant levels of the order of ±0.01°C. The thermal contact between the flat plate PHP and both the evaporator and the condenser is improved by spreading, over the heat exchange contact surfaces, a thin layer of Omegatherm® 201 thermal grease (2.3 W/m-°C). Heat is provided to the evaporator by four cartridge heaters (outer diameter 10mm and 40mm length) embedded in a copper block, thermally insulated by mineral wool. The heaters are connected to a power supply (TDK-Lambda GEN300-17).

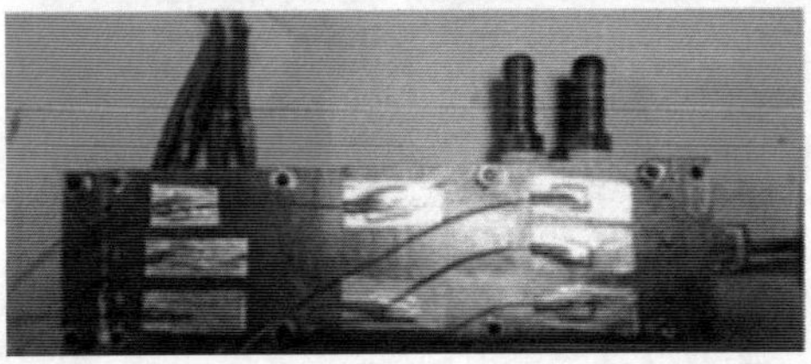

a)

b)

Figure 6. Experimental setup; a) eight thermocouples: three at evaporator zone, two at adiabatic zone and three at condenser zone; b) experimental setup including PHP, heater and heat sink.

3.1. Experimental procedure

Tests were conducted in horizontal and vertical positions bottom heat mode, for both PHP evaluated. Power levels started at 10W and reached 100W at steps of 10W. The working fluid used was 3M HFE7100 at four different filling ratios (FR) of 0, 25, 50 and 75% of the total internal volume, for each PHP. Two inclination angles (0° and 90° heating at bottom position) were also adopted as testing parameters.

4. Experimental results

4.1. Fabrication process

Microstructural analysis comparing the cross-section area of the bonded plate interfaces produced by different diffusion bonding routes, indicated a direct influence on the final microstructure, explained as follows.

4.1.1. Fabrication using a non-controlled pressure furnace. The plate interfaces of the PHP fabricated using the conventional resistive furnace (non-controlled pressure furnace), did not presented continuity of the material, as can be observed in Figure 7a and b. This microstructural aspect directly influenced the sealing capacity of the device, which was not able to maintain the necessary vacuum of 10^{-7}mbar-l/s. Furthermore, Figure 7 shows that the semi-circles did not match well to form a circular channel, due to misalignment. Although the misalignment was less than 0.2 mm, this distortion forms grooves that effect the nucleation sites during liquid-vapor working fluid phase change and meniscus formation. discontinuity between the planar bonded surfaces (Figure 7b) is due to insufficient mechanical intimacy of metal-to-metal contact, a necessary condition to obtain a satisfactory diffusion union. The pressure applied by the screws on the stainless-steel matrix device was not enough to provide a homogeneous plastic deformation of surface asperities. As a final step of the fabrication process, additional high vacuum glue was applied to the borders to keep the stagnation of the device.

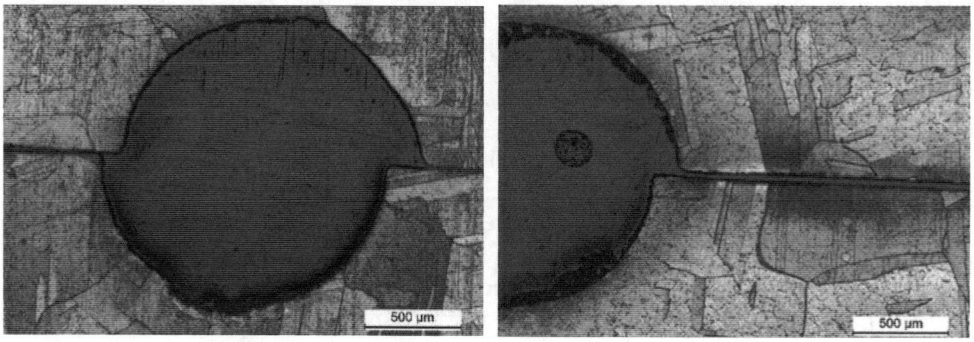

(a) (b)

Figure 7. Interface in PHP with porous media. a) Channel; b) Interface.

On the other hand, the copper powder sintered successfully to the channel surface and among particles with the used set of parameters. However, the resultant porous media was not homogeneously distributed through the channel surface, as shown in Figure 8a. This is associated with the adopted powder pouring technique, which must be optimized. Within the continuous porous media, a homogeneous porous layer can be observed, as zoomed in Figure 8b.

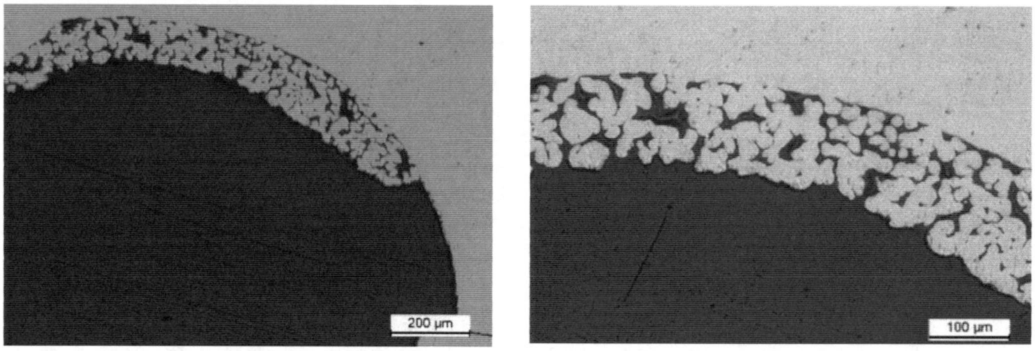

a) b)

Figure 8. Porous media distribution; a) porous media in alternating porous media PHP; b) zoom of porous media.

4.1.2. Fabrication using a controlled pressure furnace. The conventional flat plate PHP, fabricated using the high vacuum heat treatment furnace with an integrated pressing unit, presented more continuous joint interfaces. Furthermore, the plates matched better and the misalignment (gap formed between the semicircles) decreased considerably, as shown in Figure 9a. This better union quality can be attributed to both the controlled pressure, which allowed the necessary mechanical intimacy, and to the high vacuum atmosphere, which allowed the reduction of surface oxides, improving the metal to metal contact necessary for the interdiffusion of atoms through surfaces. Grain growth through the joint interface was observed in some regions (Figure 9b). Such interface continuity improvement resulted in a better vacuum capacity of the device, which could maintain the pressure of 10^{-7}mbar.l/s, without leakage.

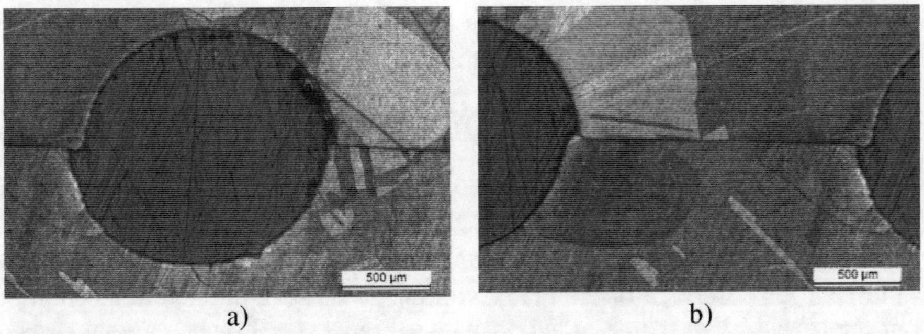

Figure 9. Cross sectional area in conventional PHP; a) Channel in conventional closed loop PHP; b) interface results in closed loop PHP.

4.2. Thermal results

Differences of the mean temperatures (ΔT) at evaporator and condenser regions are presented in Figure 10 as a function of time, where ΔT is the difference of mean temperatures at evaporator zone ($\bar{T}_e$), condenser zone($\bar{T}_c$), where:

$$\bar{T}_e = \frac{(T1 + T2 + T3)}{3} \quad (1)$$

$$\bar{T}_c = \frac{(T6 + T7 + T8)}{3} \quad (2)$$

Figure 10 shows the transient ΔT variation, start-up for conventional PHP, operating in vertical position and FR of 25 and 50%, start-up happened at power levels less than 40W. ΔT was lower for wicked PHP, FR of 25% and power input of 10 and 20W; however, ΔT increased once in the range from 30 to 40W, decreasing again at power level of 70W. Wicked PHP at vertical position and FR of 50 and 75% and smooth conventional PHP with FR 75% achieved start-up at power input of 60W.

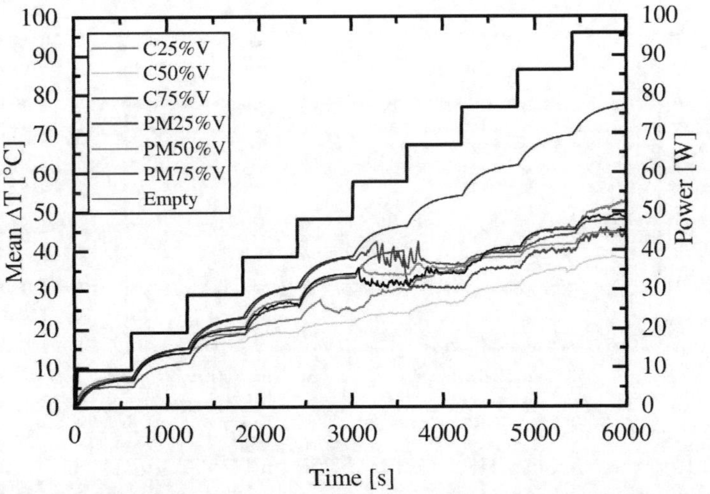

Figure 10. Differences of mean temperatures; C and PM are conventional and porous media PHP, FR of 25, 50 and 75% and vertical position (V).

However, start-up at horizontal position was not observed for all the experimental tests, although some temperature oscillations were observed, heat transfer mechanism at horizontal position is mainly the conductive one.

Using ΔT for each the thermal power supplied (Q), the thermal resistances (R_t) are defined as:

$$R_t = \frac{\Delta T}{Q} \tag{3}$$

The thermal resistances observed for the PHPs working at the several operation conditions selected in this paper are presented against heat power input in Figure 11. Lower thermal resistances were obtained for both PHPs in vertical position, as expected from Figure 10, as in this condition, the PHPs operate in thermosyphon mode. Moreover, less time was necessary for the wicked PHP and FR of 25% to achieve start-up at low power inputs; however, when the power supplied was found between 30 and 60W, thermal resistance rises for this experimental case. Conventional PHP with FR of 25 and 50% and vertical position started to work at 20W and the thermal resistance followed a decreasing trend for all the power range, achieving a thermal resistance about of 0.4°C/W and 0.5°C/W for FR of 50 and 25% respectively. Other filling ratios at vertical positions decreases the thermal resistance around of 0.5°C/W.

As told before, both devices with any filling ratios did not work at horizontal position; thus, the thermal resistance measured in this case was only due to the conduction heat transfer. Pure conduction was effectively measured for the PHP with 0% of filling ratio (void device). The overall resistance of the empty PHPs is close to the PHPs operating in the horizontal positions, showing that phase change is practically not generating oscillations. The device with alternating porous media, horizontal position and charged with FR of 25% tried to work, as one can see by some small decreasing of the thermal resistance curve. From all the configurations tested, the conventional PHP and FR of 50%, operating in vertical position presented the lowest thermal resistance. Thermal resistances uncertainty was calculated to be less than 10% at power levels from 10 to 50W and less than 2% for power inputs higher than 60W.

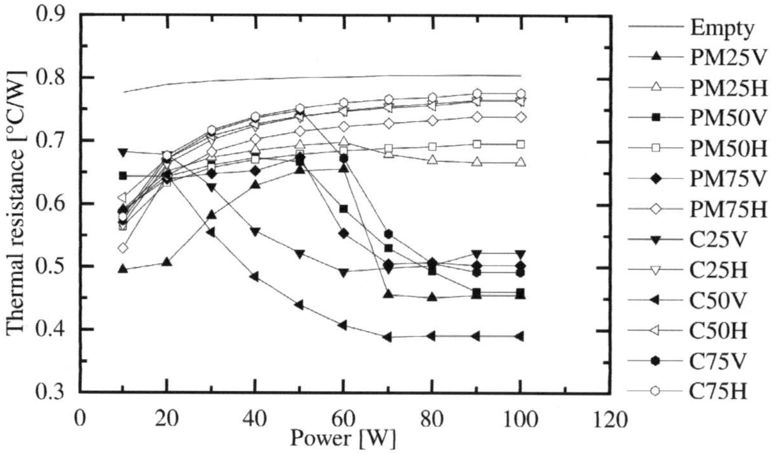

Figure 11. Thermal resistance. where C and PM are conventional and porous media PHP, FR of 25, 50 and 75% and H and V are horizontal and vertical positions.

5. Conclusions

This article presents two flat plate closed loop PHPs fabricated using the diffusion bonding technology, with and without the application of sintered loose power wick porous media over the internal surfaces of alternated evaporator channels. Both PHPs were tested under the same experimental conditions. From the present work, on can conclude that:

Microscopical investigations revealed that the application of pressure in combination with a reducing atmosphere is of utmost importance to achieve continuous joint interfaces, which directly affected the devices ability to maintain the vacuum necessary to service operation. Such analysis also indicated that the loose powder sintering is a suitable technique to produce the porous media that can recover channel surfaces of minichannels as flat plate PHPs. However, to obtain channels with a homogeneous distribution of porous layer, optimizations on the powder pouring process need to be developed.

The mismatching of the plate grooves due to assembly deviation results in deformed channels that yields capillary gaps between united plate surfaces, affecting the thermohydraulic performance of the PHPs. Further analysis needs to be focused on this phenomenon.

Porous media in alternating channels hinders the fluid motion due to pressure drops, results does not show improvements in the overall device thermal performance. Nevertheless, the wick allowed a weak early PHP start-up at filling ratio of 25% and only for power inputs up 20W.

The thermal resistance of the conventional PHP operating in vertical position decreased by almost 45% for 50% of filling ratio, when compared to the empty PHP, where no phase-change takes place. This means that the conduction heat transfer is important in this case, as expected, once the device has a considerable mass of solid material (copper), of high conductivity and high density. Further studies need to be developed, focusing on optimizing the plate geometry, aiming the decrease of the thermal resistance and thermal inertia.

6. Nomenclature

C	: conventional
CNC	: Computerized numerical control
FR	: Filling ratio [%]
H	: Horizontal
PHP	: Pulsating heat pipe
PM	: Porous media
R_t	: Thermal resistance (°C/W)
ΔT	: Temperature difference (C)
Q	: Thermal power supplied (W)
$\bar{T}_e$	: Mean T at evaporator zone (°C)
$\bar{T}_c$	: Mean T at condenser zone (°C)
V	: Vertical position

7. References

[1] Mangini D, Mameli M, Georgoulas A, Araneo L, Filippeschi S and Marengo M 2015 A pulsating heat pipe for space applications: Ground and microgravity experiments *Int. J. Therm. Sci.* **95** 53–63

[2] Charoensawan P, Khandekar S, Groll M and Terdtoon P 2003 Closed loop pulsating heat pipes - Part A: Parametric experimental investigations *Appl. Therm. Eng.* **23** 2009–20

[3] Chien K H, Lin Y T, Chen Y R, Yang K S and Wang C C 2012 A novel design of pulsating heat pipe with fewer turns applicable to all orientations *Int. J. Heat Mass Transf.* **55** 5722–8

[4] Karimi G and Culham J R 2004 Review and assessment of pulsating heat pipe mechanism for high heat flux electronic cooling *Inter Soc. Conf. Therm. Phenom.* 52–9

[5] Yang K S, Cheng Y C, Liu M C and Shyu J C 2015 Micro pulsating heat pipes with alternate microchannel widths *Appl. Therm. Eng.* **83** 131–8

[6] Tseng C Y, Yang K S, Chien K H, Jeng M S and Wang C C 2014 Investigation of the performance of pulsating heat pipe subject to uniform/alternating tube diameters *Exp. Therm. Fluid Sci.* **54** 85–92

[7] Schwartz M M 1969 *Modern metal joining techniques* (New York: John Wiley & Sons)
[8] Rusnaldy R 2001 Diffusion bonding: an advanced of material process *Rotasi* **3** 23–7
[9] Thümmler, F; Oberacker R 1993 *An introduction to powder metallurgy* (London: Institute of Materials)
[10] Vieira G, Nuernberg G, Londoño N and Mantelli M 2017 Stress analysis of a diffusion bonding device under loading and thermal expansion *24th ABCM International Congress of Mechanical Engineering* (Curitiba-Brazil)

Acknowledgement
The authors acknowledge the help of all the members of the Heat Pipe Laboratory LABTUCAL and Materials Laboratory LABMAT at the Federal University of Santa Catarina.

Flat Plate Pulsating Heat Pipes with and without separating grooves: experimental investigation on adiabatic length, coolant temperature and orientation

F Pagnoni[1,2], V Ayel[1*], C Romestant[1], Y Bertin[1]

[1] Institut PPRIME, CNRS-ENSMA-Université de Poitiers, UPR 3346, Futuroscope Chasseneuil - Cedex, 86961, France
[2] Department of Aerospace Engineering, Politecnico di Milano, Milan, 20158, Italy

Abstract. Two Flat Plate Closed Loop Pulsating Heat Pipes have been tested under various operating conditions, working fluids and different evaporator-to-condenser distances. The devices have the same geometrical features; nevertheless, on one FPPHP plate external grooves between adjacent channels have been engraved to minimize the transversal thermal conductive spreading whilst the other is a simple smooth plate. The working fluids are ethanol and FC72. Both FPPHPs have shown better performances when tilted in favorable vertical position. It has been found that higher secondary fluid temperatures within the condenser provide better thermal performances. The fluids' different thermophysical properties and critical diameters have remarkable effects on FPPHPs behavior. Moving the condenser towards evaporator increases the heat transfer capabilities, reducing the influence of gravity. Finally, the higher thermal insulation between adjacent channels, provided by the external grooves, seems to degrade the thermal performance, especially on horizontal position and on the edge.

1. Introduction

Conceived by and patented by Akachi [1] in 1990, Pulsating Heat Pipes (PHPs) are passive, two-phase heat transfer devices that have been intensively studied over the last two decades. Their cheap and simple structure which consists of one bended tube arranged in a coil shape (for classics PHPs) or in a channel engraved within a thin conductive plate (FPPHPs), makes the Pulsating Heat Pipes highly appealing solutions for some industrial purposes, in particular electronic cooling with high heat flux densities and reduced scale heat exchangers [2]. However, their chaotic and pulsating flow nature makes the thermo-hydraulic coupling difficult to predict numerically and experimentally.

Nevertheless, important steps forward have been done both from experimental [3] and numerical [4] point of views. Indeed, visualization campaigns like those carried out by Khandekar *et al.* [5] and more recently by Xue *et al.* [6] allowed identifying the main flow regimes within the PHP: slug, semi-annular and annular ones. Furthermore, literature provides a number of parametric studies on Pulsating Heat Pipes that attempt to correlate geometrical features, operative conditions to the thermal performance.

* Corresponding author: vincent.ayel@ensma.fr

These studies are of primary importance because they can outline some key features of the PHPs, giving helpful guidelines for their design and addressing future researches in the field.

For example, Gi *et al.* [7] analyzed the influence of filling ratio on thermal performance: they found that, depending on the working fluid, the best occurred for values within the range 25-65%. Charoensawan *et al.* [8] studied the effects of the channels number on the PHP behavior: they argued that a minimum number of 16 channels is able to reduce the effects of gravity and ensures the functioning of the PHP in horizontal position, for both internal diameters of 1 and 2 mm. Mameli *et al.* [9] correlated the PHP inclination with respect to gravity to thermal performance at different heat inputs. Bonnenfant *et al.* [10] compared open and closed loop PHP configurations: they found that at low input heat powers, the latter provides a better thermal performance compared to the former.

Ayel *et al.* [11] investigated the behavior of a FPPHP in a variable gravity field: they observed a quick thermal response during the gravity module variation and a satisfactory thermal performance on horizontal position compared to the one obtained on ground tests.

According to different authors [11-12], in FPPHP made of full plates, the thermal inertia of the latter generally prevents the device from experiencing abrupt temperature peaks during operation. Moreover, conduction thermal spreading occurs through the full plates. As a result, lower temperature gradients are generally observed as well as pressure homogenization between the channels. Since in a PHP pressure differences among adjacent channels are the main driver of oscillations under slug flow, FPPHP are potentially more exposed to a premature dry-out in horizontal inclination.

Since this latter aspect has been rarely investigated in the literature, the present paper reports the results of a multi-parametric study performed on a special FPPHP having external grooves machined on one plate to partially separate internal channels. The present work reports a quantitative analysis of the PHP steady-state thermal performance as function of the heat power inputs, heat sink temperature, orientation with respect to gravity and evaporator-to-condenser lengths. Finally, a comparison with a full plate FPPHP that has the same geometrical dimensions is also provided.

Most of the results presented in this work agree with those available in the literature. Nevertheless, some divergences emerged, underlying, once again, that the complex thermohydraulic behavior of PHPs is still far from being fully understood.

2. Experimental setup

2.1. Closed Loop Flat Plate PHPs

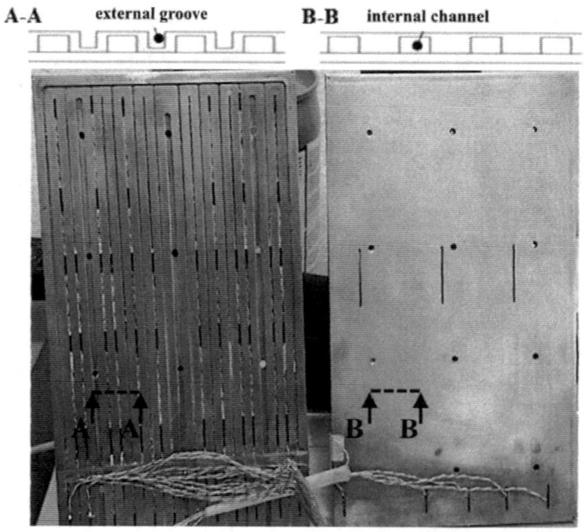

Figure 1. Front view (bottom) and schematic channel cross section (top) of the FPPHPs: with external grooves 'GP' (left), smooth plate 'SP' (right).

Two closed loop Flat-Plate Pulsating Heat Pipes have been tested under various conditions. The FPPHPs have been manufactured from a 120x200x2 mm^3 copper plate in which square grooves of 1.6x1.7 mm^2 have been machined to obtain 24 parallel channels (23 U-turns). Channels are then closed by means of a second plate (1 mm thick) that has been brazed on top of the latter. The brazing operation provides a perfect sealing between the adjacent internal channels and from the external environment. External grooves of 1x1.6 mm^2 have been machined on one FPPHP to increase the transversal thermal resistance between adjacent channels, as shown in figure 1. For the sake of simplicity, as the presence of external grooves on one FPPHP represents the unique geometrical difference between the two devices, these will be named, from now on, as Grooved-Plate (GP) and Smooth-Plate (SP) -PHPs, respectively. The former device has been previously tested in a variable gravity field during the 60th ESA parabolic flight campaign.

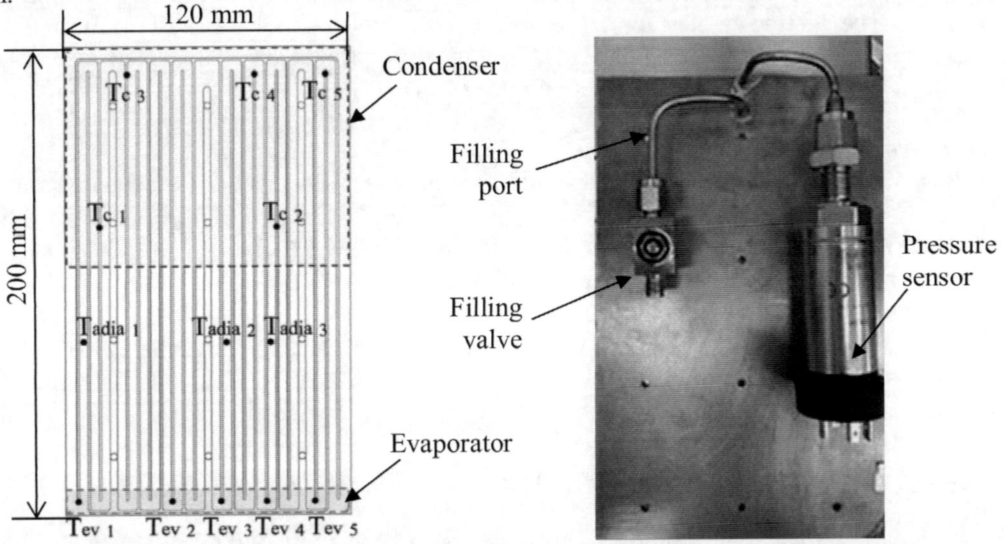

Figure 2. PHP instrumentation: thermocouples arrangement (left) and stainless-steel pipes connections for the filling port and the pressure sensor (right).

Indeed, the GP-PHP internal channel dimensions have been sized for microgravity conditions through the Weber criterion, in the range of flow speeds between 0.1-0.2 m.s^{-1}, FC72 as working fluid and temperatures up to 60°C, more details can be found in [11]. All the tests performed in this work has been conducted on ground, it follows that the internal channel dimensions should be sized following the critical Bond number, as shown in Eq. (1). As with FC72, the hydraulic diameter of the internal channel exceeds the critical one (equation (1)) for temperatures above 30°C, ethanol has been also selected as primary working fluid (d_{crit}(100°C) ≈ 3 mm).

$$d_{crit} \sim \left(\frac{\sigma}{(\rho_l-\rho_v)g}\right)^{1/2} \tag{1}$$

2.2. Test apparatus

The FPPHP evaporator consists of an electrical wire (Thermocoax, 1 mm external diameter, R = 3.81 Ω) embedded in the grooves of a thin copper plate of 10x120x2 mm^3 fixed on one edge of the FPPHP using clamps (figure 2 (left)). A controlled heat power is delivered by means of a power supply (EA® 8360-10T).

The heat sink consists of an engraved aluminum cold plate of 80x120x10 mm^3 in which a mixture of water/glycol (5% in volume) flows at a controlled bath temperature, by means of a high precision thermoregulation device (HUBER® CC240wl). The cold plate is screwed on the FPPHP plate. To enhance the thermal contact between the hot source, the heat sink and the FPPHP plate, a thermal gap

IOP Conf. Series: Materials Science and Engineering 1139 (2021) 012005 doi:10.1088/1757-899X/1139/1/012005

filler (Tflex Series Lard Technologies®, thickness δ = 0.5 mm, $\lambda \sim 6$ W.m^{-1}K^{-1}) and a thin layer of silicon paste (Heat Sink Compound, $\lambda = 2.9$ W.m^{-1}K^{-1}) have been applied at their interfaces, respectively.

Thirteen T-type thermocouples of 0.5 mm diameter have been glued on the plate to measure the PHP temperatures: five thermocouples in some grooves on the back side of the evaporator ($T_{ev\,1}$ to $T_{ev\,5}$), five within grooves on the back side of the condenser ($T_{c\,1}$ to $T_{c\,5}$) and the last three in the adiabatic region between evaporator and condenser ($T_{adia\,1}$ to $T_{adia\,3}$), as shown in figure 2. A pressure sensor (GE PTX5076, 5 bar absolutes, 0.04% accuracy on full scale) has been connected to FPPHP through a stainless-steel pipe brazed on the plate outer groove at the condenser end. The devices have been thermally insulated during the tests with insulating Rockwool (λ = 0.04 W.m^{-1}K^{-1}) and an aluminum tape coating. The thermocouples, pressure sensor and power supply are all connected to a data acquisition system (NI-cRIO-9073, NI-9214). Data recording frequencies have been set to 1 Hz for the temperature signals and 100 Hz for pressure ones.

2.3. *Test campaign*
Both FPPHPs have been tested under the operative conditions resumed below (see figure 3 as reference):

- Working fluids: FC72, ethanol (filling ratio $FR \approx 50\%$ at 20°C);
- Heat power applied: from 20 W to 260 W by steps of 30 W;
- Coolant temperatures (T_{II}): 5°C, 20°C, and 40°C;
- Orientations: horizontal (α = 0°) with gravity vector perpendicular to the FPPHP plate, with a favorable tilting angle (α = 45°) with respect to ground reference, vertical favorable (α = 90°) and on the edge, which corresponds to the case of α = 0° with gravity vector parallel to the FPPHP plate;
- Different evaporator-to-condenser distances: L_1 = 10.8 cm, L_2 = 8.7 cm and L_3 = 5.8 cm.

For each of these operating conditions, heat power has been applied to evaporator by steps of 30 W starting from 20 W up to 260 W or less if the evaporator's dry-out limit was encountered and/or temperatures exceeded 120°C. Each power step has been maintained until the FPPHP reached a relatively steady state mode for both temperatures and pressure signals.

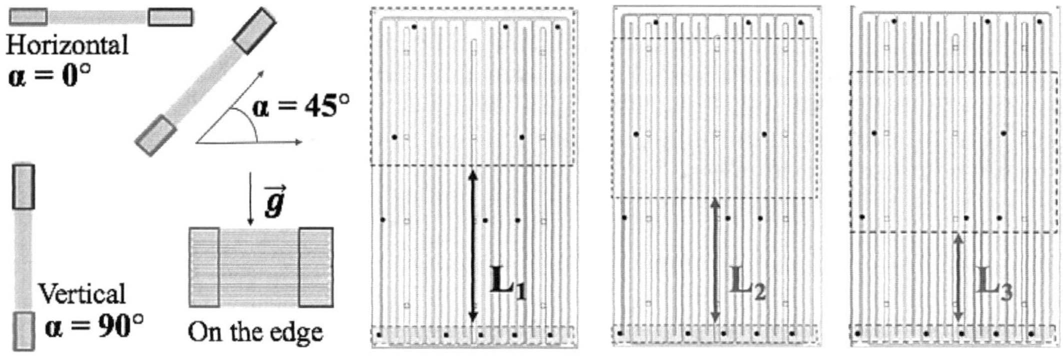

Figure 3. Illustration of the various FPPHP operating configurations tested: Orientation with respect to gravity (left) and different adiabatic lengths (right).

2.4. *Data processing*
The steady state thermal performances at each heat power step has been quantified through the overall thermal resistance, computed as in equation (2); the time and space averaged temperatures of evaporator and condenser, respectively, over the net heat power applied to the evaporator. For the latter, an

estimation of the heat leaks have been performed using the data from the empty PHP test; assuming a two resistance model and minimizing the sum of the mean squared deviation of the computed and measured input heat powers, it was possible to find the thermal resistances of the empty PHP (pure conductive) and the one corresponding to the heat losses with the external environment.

$$R_{th} = \frac{\bar{T}_{ev} - \bar{T}_c}{Q_{in} - Q_{loss}} \tag{2}$$

An uncertainty analysis has been performed for both temperatures and the measured input heat powers, in order to estimate the impact of the systematic uncertainties on the thermal resistance values. Assuming, for small perturbations, a linear relation between the input quantities and the observed variable, the propagation of the uncertainties for the thermal resistance can be estimated as shown in equation (3).

$$e_{R_{th}} = \pm\left(\left(\frac{\partial R_{th}}{\partial Q}\right)^2 e_Q^2 + \left(\frac{\partial R_{th}}{\partial T}\right)^2 e_T^2\right)^{1/2} \tag{3}$$

In equation (3), the uncertainties associated to the heat powers and temperature measurements that appear within the square root can be estimated from the accuracy and stability data of both the power supply and thermocouples through an equivalent expression of equation (3). In the range of temperatures and heat powers supplied, the systematic uncertainties for thermal resistances are in the range ± 0.02 KW^{-1}. It has been seen that the standard deviations of temperatures around their time-averaged values have higher impact on the thermal resistances rather than the systematic uncertainties, as argued by Ayel *et al.* [11].

All tests were performed in a period of four months, the devices have been stored in temperature-controlled room and the FPPHPs have shown a satisfactory level repeatability in terms of overall temperatures and pressure trends.

3. Experimental results

3.1. Effects of PHP orientation

Results for SP and GP –PHPs with ethanol and FC72 are presented in figures 4 and 5, respectively. According to Mameli *et al.* [9], three different operating zones can be distinguished: the "Start-up", "Normal operating" and "Medium High heat inputs" ones. The first zone corresponds to the low heat input levels (< 80 W) where the gravity head seems not to affect the thermal resistances that are generally decreasing as the heat inputs increase in all orientations. On the other hand, in the normal operating zone (between 80 and 200 W in figure 4, and 80 and 140 W in figure 5), a higher favorable inclination angle brings a lower thermal resistance. Here, the trends of R_{th} are still slightly decreasing or almost constants over the input heat powers.

Compared to the empty case, for the SP-PHP with ethanol (see figure 4), the values of thermal resistances in the normal operating zone decrease down to 64% in horizontal orientation, 69% on the edge, 78% at $\alpha = 45°$ and 82% in vertical one. Finally, in the "Medium high thermal inputs" zone (from 200 W in figure 4), evaporator dry-out occurs in horizontal and on the edge orientations, whilst for $\alpha = 45°$ and 90° the SP-PHP provides a continuous operation up to the maximum heat power step with no thermal crisis observed. As shown in figure 5, higher inclination angles postpone the thermal crisis, which leads to an increase of the thermal resistance (for $\alpha = 45°$) followed by a dried out evaporator condition.

A favorable hydrostatic pressure head (bottom heated mode device) enhances the PHP thermal performance. Moreover, the presence of sharp edges in the inner channel section leads to supplementary capillary forces; these tend to break up the liquid menisci and to promote an annular flow regime, like in a grooved heat pipe. Under such operating condition, the PHP reaches the highest thermal performances [12], and the highest heat power tested (260 W) does not here correspond to the thermal crisis of the FPPHP.

When oriented on the edge, the values of R_{th} are generally smaller compared to those obtained in horizontal: this belongs to the transversal hydrostatic pressure head across the liquid menisci within the interconnected channels [13]. With ethanol as working fluid, this pressure head is in the order of 30 Pa for adjacent channels and 800 Pa for the channels at the PHP edges, whilst for FC72 they are of 70 Pa and 1800 Pa, respectively. With FC72 as working fluid, both FPPHP have provided a sustained and stable functioning only when gravity assisted, but not in horizontal orientation, as shown in figure 5 for the GP-PHP at T_{II} = 20°C: this is due to the inner channel dimensions that exceeds the critical ones (see Eq. (1)) above 30°C.

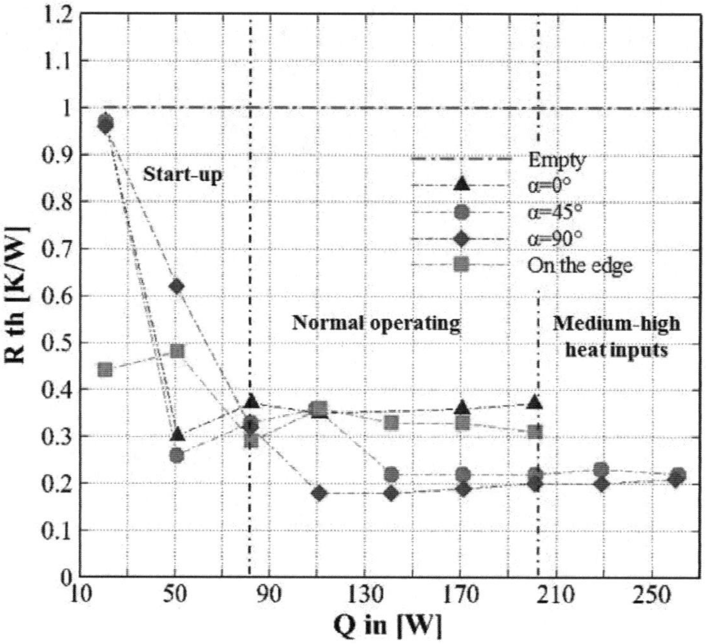

Figure 4. Effects of the PHP orientation with respect to gravity (GP-PHP, FC72, T_{II} = 20°C).

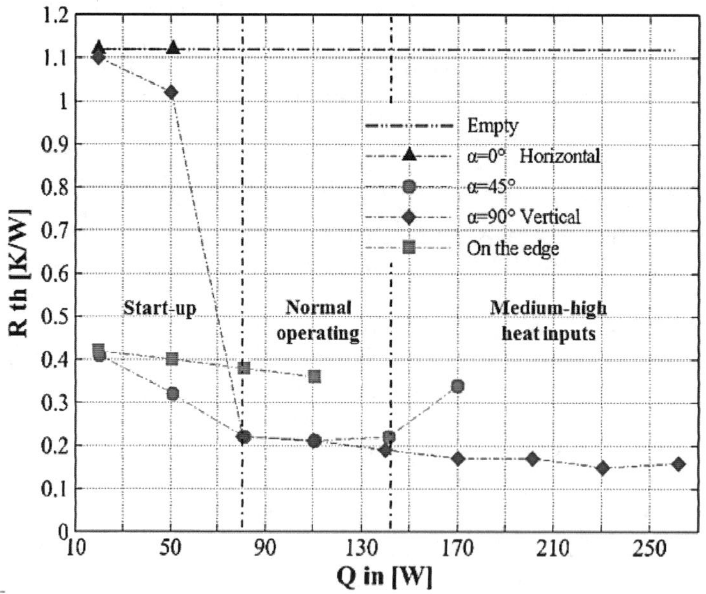

Figure 5. Effects of the PHP orientation with respect to gravity (SP-PHP, ethanol, T_{II} = 20°C).

3.2. Effects of the coolant temperature

The use of a cold plate as heat sink with a controlled coolant bath temperature allows precisely defining the minimum PHP working temperature. As outlined in figure 6, which shows R_{th} values for the GP-PHP on the edge with ethanol and FC72 as working fluids, a higher secondary fluid temperature brings lower thermal resistances. This trend has been observed for all input heat powers (with only few exceptions within the Start-up zone), all PHP orientations and both working fluids. It is obvious that the temperature-dependence of the fluid thermophysical properties can remarkably affect the PHP thermal performance.

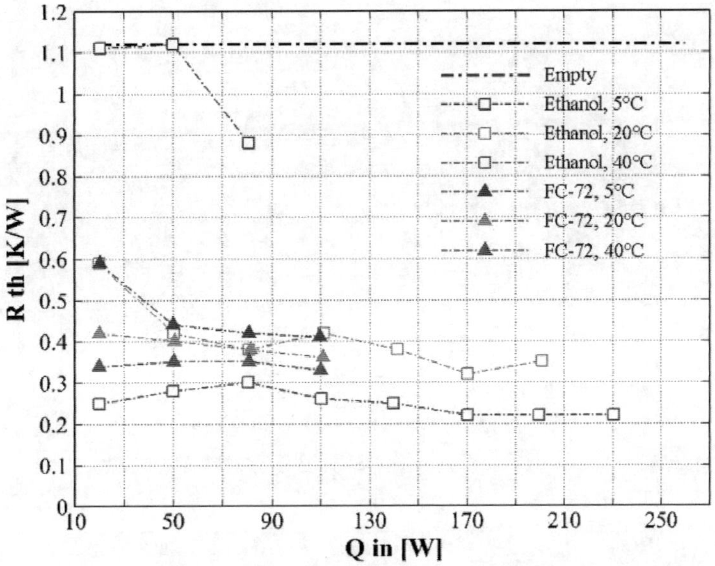

Figure 6. Effects of T_{II} on the thermal performance with ethanol and FC72 as working fluids (GP-PHP, on the edge orientation).

Indeed, switching from 5°C to 40°C, liquid dynamic viscosity of ethanol decreases of about 50% and 38% for FC72. Obviously, this means that viscous pressure losses are also decreased and the PHP performs better. However, it is interesting to observe that with FC72, R_{th} change rate is smaller compared to ethanol. This could be due to the different temperature-variation rate of the thermophysical properties between the two fluids, as shown in Table 1, but also to a different flow regime induced by the internal channel dimensions with respect to the critical ones.

Table 1. Main thermophysical properties of ethanol (left column) and FC72 (right column) between 5˚C and 100˚C.

T_{sat} (˚C)	P_{sat} (kPa)		h_{lv} (kJ.kg^{-1})		ρ_l (kg.m^{-3})		$C_{p,l}$ (kJ.kg^{-1}K^{-1})		μ_l (mPa s)		λ_l (W.m^{-1}K^{-1})		σ_l (mN.m^{-1})	
5	2.35	11.3	1044	98	802	1732	2.23	1.02	1.65	0.86	0.182	0.058	24	14
20	5.8	23.6	1030	94.2	790	1702	2.21	1.04	1.2	0.70	0.181	0.056	23	12
40	18	54.8	1012	88.9	772	1660	2.58	1.07	0.82	0.54	0.175	0.054	21	10
60	47.2	112.4	989	83.6	754	1614	2.79	1.10	0.59	0.43	0.171	0.052	19	8
80	109	208.7	960	77.8	735	1560	3.04	1.14	0.43	0.36	0.169	0.049	17	6
100	226	358.5	927	71.4	715	1490	3.31	1.17	0.32	0.31	0.167	0.047	16	5

3.3. *Effects of the evaporator-to-condenser distance*

Evaporator-to-condenser distance represents the heat transport length of the PHP (generally called adiabatic zone in classic heat pipes); its variation implies also a change of the pressure losses generated in the flow. The results presented in figure 7 concern the GP-PHP with ethanol as working fluid, T_{II} = 20°C and different evaporator-to-condenser distances (see figure 3): L_1 (corresponding to a classic configuration with condenser at top end of the plate), L_2/L_1 = 0.81 and L_3/L_1 = 0.54. Obviously, moving the cold plate towards evaporator has generated a secondary adiabatic region that, differently from the primary one, communicates only with condenser, as shown in figure 3. The temperatures recorded in this area have been considered for the estimation of the averaged condenser ones in equation (2).

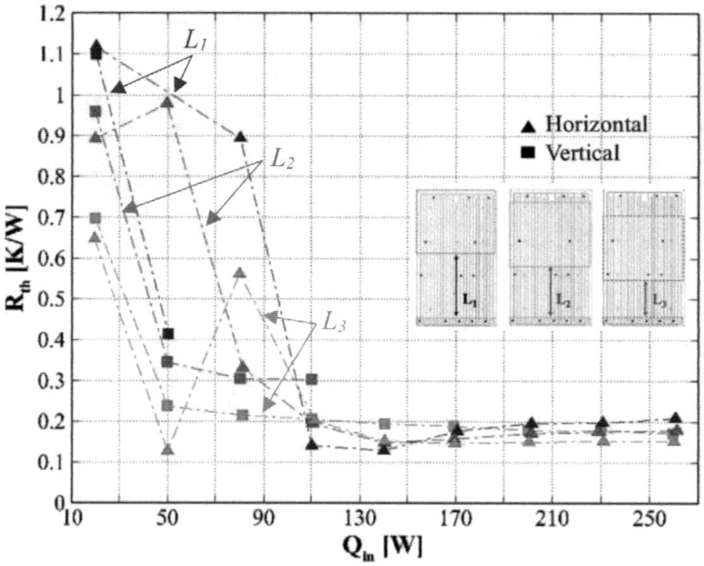

Figure 7. Effects of evaporator-condenser distance in horizontal and vertical position (GP-PHP, ethanol, T_{II} = 20°C).

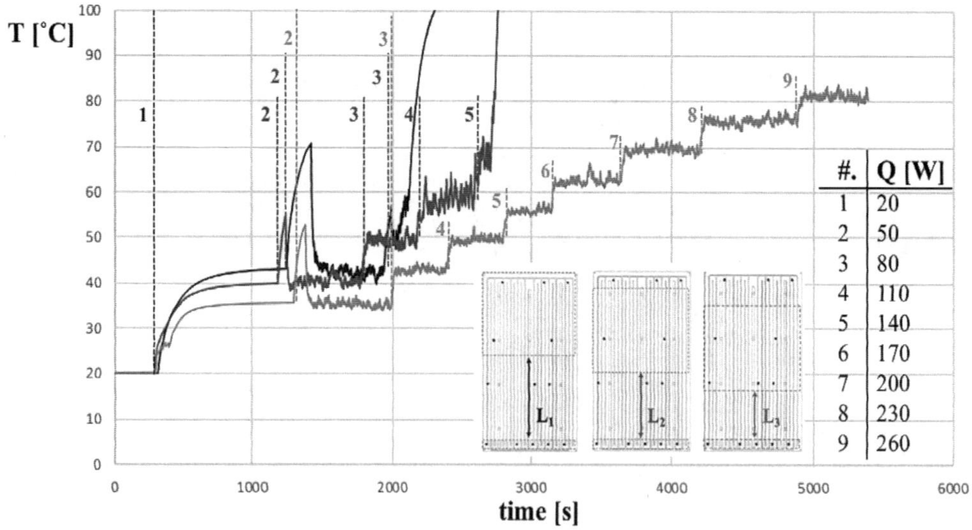

Figure 8. Averaged evaporator temperatures for different evaporator-condenser distances (GP-PHP, ethanol, horizontal, T_{II} = 20°C).

As shown in figure 7, adiabatic length plays an important role on the PHP behavior and performances: thermal resistances are decreased when condenser is moved towards evaporator.

Furthermore, with the minimal distance L_3, gravity seems not to affect the thermal performance anymore. Indeed, the gap between the horizontal and vertical positions almost disappear and all curves are well superimposed in the normal operating region ($Q_{in} \geq 90$W). Looking at space-averaged evaporator temperature curves for the three distances in figure 8, the GP-PHP oscillation start-up occurred always at 50 W (step #2) with an abrupt temperature drop. Then, a decrease in the heat transport length postpones the evaporator dry-out and provides lower temperatures for the same heat inputs. Similar results have been obtained with FC72.

3.4. Effects of the separating grooves

The presence of external inter-channel grooves on GP-PHP (see figure 9) increases the transversal conductive thermal resistance of about 65% compared to the SP-PHP. As argued by Khandekar *et al.* [12], the inter-channel heat balance plays a negative role in PHP operation because it reduces the local temperature gradient, which causes a decrease of the pressure instabilities degrading the thermal performance. As shown in figure 10, using ethanol as working fluid, the GP-PHP provides higher temperature difference between the middle and the right-hand points of the evaporator compared to SP-PHP; starting from 140 W they increase of almost 220%. A similar trend has been observed between other couples of points within the evaporator. However, figure 11 outlines clearly that the overall thermal performance of GP-PHP is degraded compared to SP-PHP one. Indeed, on the edge orientation and T_{II} = 20°C, the R_{th} values of the former are between 10% and 30% higher between 80 and 140 W, closer values are recorded at higher heat power steps.

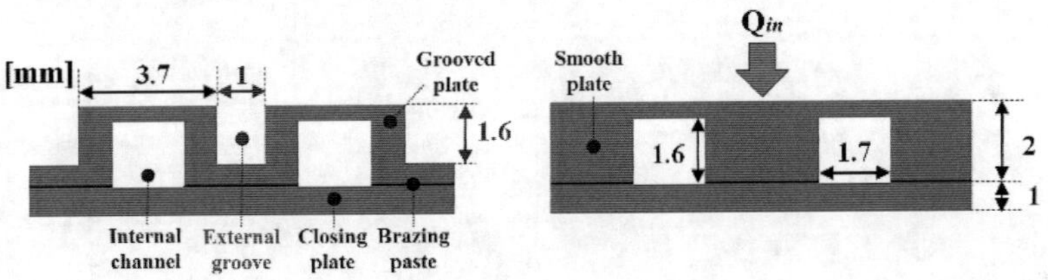

Figure 9. Schematic geometry of GP and SP –PHP channel cross-sections.

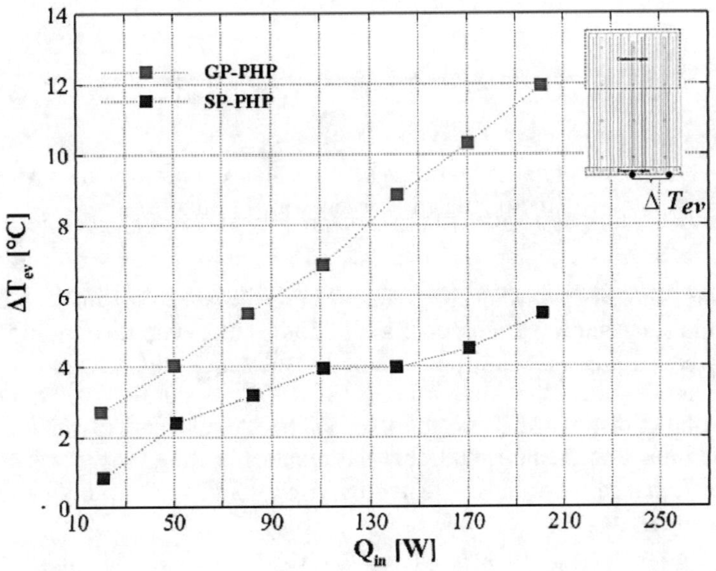

Figure 10. Effect of external grooves on the time-averaged, steady state, evaporator's temperature differences in two points (ethanol, on the edge, T_{II} = 20°C).

Moreover, in horizontal position, the GP-PHP evaporator was totally dried-out and its performances were close to those of the empty case, whilst the SP-PHP provided a steady thermal performance keeping evaporator's temperatures below 120°C at more than 140 W. Using FC72, R_{th} values are well superimposed and both PHPs have a dried-out evaporator condition in horizontal position. The results with ethanol are meaningful because if, on one side, the presence of external grooves enhances flow instabilities, on the other hand they seem to limit the heat transport capability of the FP-PHP, probably because of the less homogeneous heat flux distribution around each channel with higher peaks in the grooved plate outer side, where the heat power is applied.

This could probably cause the sudden formation of a dry spot starting from low input heat powers that prevent liquid from rewetting the evaporator de-priming flow oscillations. When oriented on the edge, the situation improves for the GP-PHP as long as flow instabilities are also sustained by the hydrostatic pressure head.

This trend appears to be surprising because, in horizontal orientation, an increase of instabilities between channels, provided by the presence of the external grooves, should have improved performances. It appears that it was not the case during these experiments and this specific point deserves to be further studied.

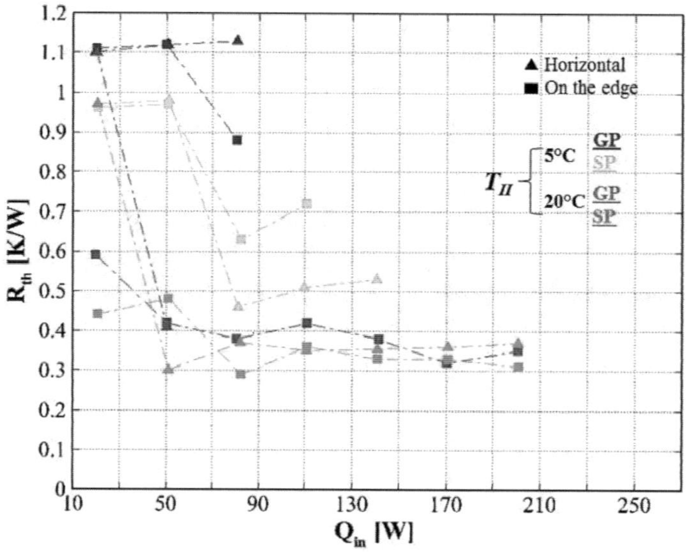

Figure 11. Effects of external grooves on the thermal performance in horizontal and on the edge orientations (ethanol).

4. Conclusions

Two copper Flat Plate, Closed Loop Pulsating Heat Pipes with a 1.6x1.7 mm^2 channels cross section have been tested on ground under various operating conditions using ethanol and FC-72 as working fluids with an equivalent filling ratio of 50%. External insulating grooves have been machined on one PHP (GP-PHP) whilst the other is a smooth plate (SP-PHP). According to Mameli et al. [9], three different zones have been distinguished on the thermal resistance curves over the input heat power: Start-up, Normal operating and Medium-high thermal inputs.

- The thermal performance is strongly affected by gravity and vertical orientation always provides the lowest thermal resistances.
- The secondary fluid temperature also has a strong impact due to the temperature-dependence of the thermophysical properties of the primary working fluid.

- It has been found that increasing the coolant temperature provides lower thermal resistances due to the decrease of the liquid dynamic viscosity of both ethanol and FC72.
- Moving condenser towards evaporator the thermal performance increases, the influence of gravity is reduced and evaporator dry out is postponed.
- Finally, the presence of insulating grooves on the external side of the PHP plate seems to not improve the performances. Nevertheless, temperature oscillations are increased up to two times, showing a good fluid flow activation during steady-state operation.

5. Nomenclature

α	Inclination angle (deg)
λ	Thermal conductivity ($W.m^{-1}.K^{-1}$)
δ	Thickness (mm)
σ	Surface tension ($N.m^{-1}$)
ρ	Density ($kg.m^{-3}$)
d	Diameter (m)
e	Uncertainty ($K.W^{-1}$) – (W) – (K)
g	Gravity acceleration module ($m.s^{-2}$)
L	Length (m)
Q	Heat power (W)
R_{th}	Thermal resistance ($K.W^{-1}$)
R	Electrical resistance (Ω)
T	Temperature (°C)

Subscripts

adia	adiabatic
c	condenser
crit	critical
e	evaporator
in	input
l	liquid
loss	losses
v	vapor
II	secondary fluid

6. References

[1] Akachi H 1990 US Patent, Patent No. 4921041.
[2] Sreenivasa R, Gupta A K and Rama N 2016 *J. Appl. Fl. Mech.* **9** 615-624.
[3] Marengo M and Nikolayev V 2018 Pulsating Heat Pipes: Experimental Analysis, Design and Applications. *Encyclopedia of Two-Phase Heat Transfer and Flow IV* vol. 1, ed J R Thome (World Scientific) 1-62.
[4] Nikolayev V and Marengo M 2018 Pulsating Heat Pipes: Basics of Functioning and Numerical Modeling, *Encyclopedia of Two-Phase Heat Transfer and Flow IV* vol. 1, ed J R Thome (World Scientific) 63-139.
[5] Khandekar S, Charoensawan P, Groll M, and Terdtoon P 2003 *Appl. Therm. Eng.* **23** 2021-2033.
[6] Xue Z, Qu W and Xie M 2013 *Prop. and Power Res.* **4** 263-268.
[7] Gi K, Sato F and Maezawa S 1999 *Proc. 11th Int. Heat Pipe Conf. (Tokyo, Japan)* 373–377.
[8] Charoensawan P and Terdtoon P 2008 *Appl. Therm. Eng.* **28** 460-466.
[9] Mameli M, Marengo M, Filippeschi S and Manno V 2013 *Proc. 17th Int. Heat Pipe Conf. (Kanpur, India).*

[10] Bonnenfant J F, Bertin Y, Ayel V and Romestant C 2012 *Proc.* 16th *Int. Heat Pipe Conf. (Lyon, France)*.
[11] Ayel V, Araneo L, Scalambra A, Mameli M, Romestant C, Piteau A, Marengo M, Filippeschi S and Bertin Y 2015 *Int. J. Therm. Sc.* **96** 23-34.
[12] Khandekar S 2003 *Microsc. Thermophys. Eng.* **6** 303-317
[13] Manno V, Filippeschi S, Mameli M, Ayel V, Romestant C and Bertin Y 2015 *Int. Phen. Heat Tr.* **3** 413-425.

Cooling of electronic components on the LTCC module with an embedded flat heat pipe

D A Nesterov[1]*, **V A Derevyanko**[1], **S B Suntsov**[2]

[1] Institute of Computational Modelling of the Siberian Branch of the Russian Academy of Sciences, 660036, Krasnoyarsk, Russia,
[2] Joint Stock Company "Academician M.F. Reshetnev "Information Satellite Systems", Zheleznogorsk, Krasnoyarsk Region, Russia

Abstract. The results of the experimental investigations of a ceramic circuit board sample with an embedded flat heat pipe are presented. The sample was made based on the technology of low temperature co-fired ceramic (LTCC). The flat heat pipe is formed at the bottom side of a ceramic substrate using a capillary structure made of sintered copper powder. The developed prototype of the LTCC module was used for the investigation of removing and spreading of high heat fluxes from high power electronic components mounted on the module. The embedded flat heat pipe filled with acetone provides 40 W/cm^2 of heat flux removal from the heat source with the heat dissipation area of 1 cm^2. The measured value of the thermal resistance between the heat source and cooler surface under the module is 0.8 °C/W.

1. Introduction

The Low Temperature Co-fired Ceramic (LTCC) technology is widely used for manufacturing electronic modules due to excellent high-frequency dielectric properties and the possibility of three-dimensional integration of passive electronic components inside one multilayer substrate [1, 2]. The thermal conductivity of LTCC is not high (2-3 W/m·K), therefore cooling of the high-power miniature components mounted on the LTCC module is a relevant problem especially under vacuum conditions. One of the solutions is the integration of a flat heat pipe directly into the LTCC module [3, 4, 5]. The flat heat pipe allows effectively spreading heat from miniature high-power heat sources, which significantly reduces the density of heat fluxes and facilitates further heat removal from the module [6].

This paper presents the results of the experimental investigation of a ceramic circuit board test sample with an embedded flat heat pipe. The sample is designed for investigating the possibilities to increase the power of electronic components based on the LTCC modules. The developed prototype of the LTCC module includes a flat heat pipe, which is intended for removing and spreading the concentrated heat load from the high-power electronic components mounted on the module. The spreading of the high heat flux over the entire area of the flat heat pipe allows further removing the heat from the bottom side of the module with a lower heat flux and consequently, with a smaller temperature drop.

* corresponding author: ndanda@icm.krasn.ru

2. Description of the problem

The development of the LTCC-module sample was based on the following main requirements:

- The module contains miniature electronic components with the dissipated heat flux $q_s \approx 20$ W/cm^2;
- The heat must be removed from the bottom surface of the module, which is mounted on a heat-removing base with a thermal grease or adhesive having a low thermal conductivity value;
- The heat-removing base under the module has the limit for the maximum heat flux value of about ≤ 3 W/cm^2;
- The module with high-power components should be a self-contained unit which can easily be installed or replaced in electronic equipment.

The high-power heat sources in the module are active electronic components that are installed on the top surface of the ceramic board. At the considered input heat fluxes of $q_s = 20$ W/cm^2, the heat transfer to the bottom surface of LTCC with the thickness of $d \approx$ 1-2 mm leads to the temperature drop $\Delta T \sim 100$ °C. The well-known solution to increase the effective thermal conductivity of LTCC in the transversal direction is thermal vias embedded into the LTCC layers under the heat sources [7]. The thermal vias allow reducing the temperature drop across the ceramic layers by 4-6 times, but the problem of further high heat flux removal from the bottom surface of the LTCC module remains. At the considered high heat flux, the temperature drop on the thermal contact between the module and the heat-removing base, on which the module is installed, is up to tens of degrees. Moreover, the cooling system under the module may not be able to remove such high heat fluxes.

The solution of the problem requires reducing the high heat flux. Flat heat pipes (vapor spreaders) are efficient thermal devices that can reduce the local input heat flux by spreading it to a large area with a very small temperature gradient. The use of the flat heat pipe in the LTCC-module can spread the heat flux from the miniature high-power component with the area of $\approx$1 cm^2 to the entire module area of 25 cm^2. This will reduce the heat flux by an order of magnitude. Since the module must be a self-contained unit which can easily be installed or replaced in electronic equipment, the flat heat pipe must be embedded into the module. The transfer of the high heat flux from the heat sources to the flat heat pipe (to the vapor in the heat pipe) should be maximum effective.

3. Design of the sample

The sample is designed for investigating the possibilities to increase the power of electronic components based on the LTCC modules. The developed prototype of the module consists of a LTCC substrate of 50 x 50 mm, which is combined constructively with a flat heat pipe (figure 1).

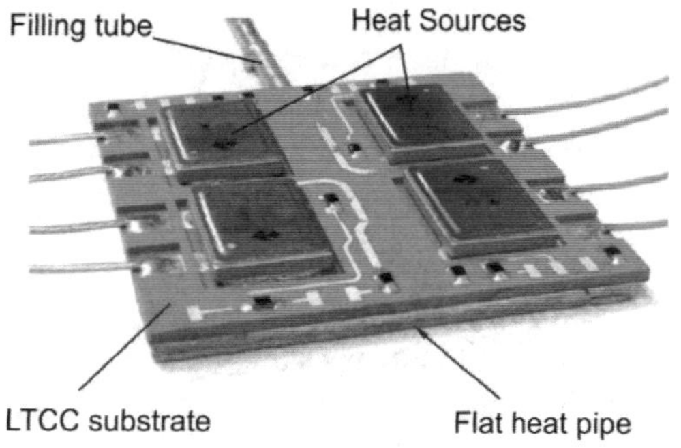

Figure 1. LTCC-module with the flat heat pipe.

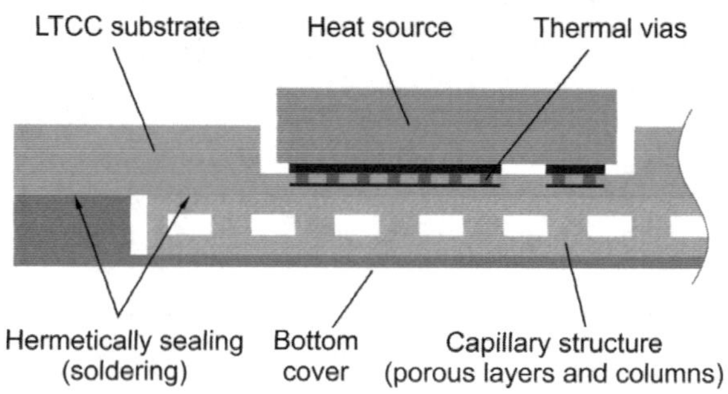

Figure 2. Internal structure of the test sample.

The capillary structure of the flat heat pipe is made of copper sintered powder. It consists of two porous layers of 0.25 mm each (figure 2). Between the layers there is an array of 11 × 11 (121 in total) cylindrical porous columns arranged with the step of 4 mm. The columns have a height of 1 mm and diameter of 2 mm. The space between the porous layers forms a vapor core. The capillary structure is attached to the lower metalized surface of the LTCC substrate and it is covered by a bottom cover, which has a hermetically sealed contact with the LTCC substrate along the perimeter. The bottom cover is made of kovar (nickel-cobalt ferrous alloy), which is compatible with the thermal expansion characteristics of LTCC. The bottom cover and the LTCC substrate form the case of the flat heat pipe.

The flat heat pipe was filled with 0.95 g of acetone (1.2 cm^3 at 20 °C). The optimum amount of the working fluid was determined experimentally and it exceeded the estimated total pore volume of the capillary structure of 0.8 cm^3. Acetone was used due to the need to operate the heat pipe at low temperatures (down to minus 40 °C). Methanol can also be considered as a promising working fluid, especially for higher operating temperatures of the heat pipe. Additionally, it can be noted that water is not suitable for the heat pipe design used. An attempt was made to use water, but the temperature difference along the heat pipe was too high. Apparently, kovar is incompatible with water. Thus, only the results for acetone are presented here.

4. Heat transfer in the sample

The heat transfer scheme is shown in figure 3. The module is mounted on a heat-removing base (cooler) with the temperature T_C. The high heat flux q_s from the heat source (transistor) on the module is transferred through solder layers, ceramic layers and a porous wick to the evaporation zone of the working fluid in the flat heat pipe. The vapor spreads over the entire area of the heat pipe and condenses on the lower layer of the wick. Then, the low heat flux q_c is transferred through the wick layer, bottom cover of kovar and thermal grease to the cooler.

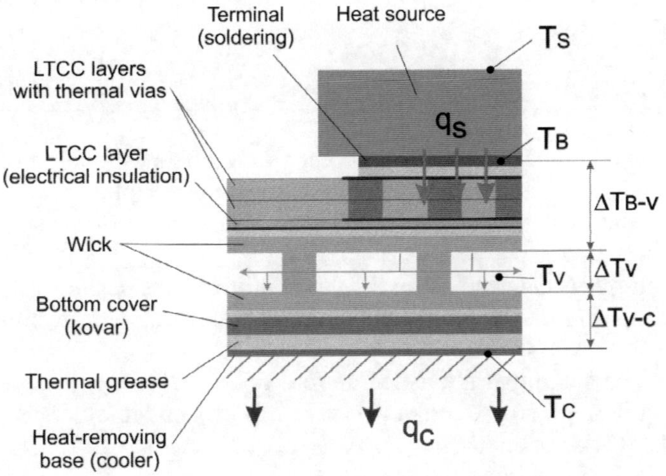

Figure 3. Heat transfer in the sample.

The temperature drop between the terminal of the transistor T_B and the cooler T_C can be expressed as a sum of three terms:

$$\Delta T_{B\text{-}C} = \Delta T_{B\text{-}V} + \Delta T_V + \Delta T_{V\text{-}C}, \tag{1}$$

where $\Delta T_{B\text{-}V}$ is the temperature drop between the terminal of the transistor and the vapor in the heat pipe, ΔT_V is the temperature drop along the moving vapor in the flat heat pipe, $\Delta T_{V\text{-}C}$ is the temperature drop between the vapor and the surface of the cooler.

The temperature drop $\Delta T_{B\text{-}V}$ occurs at the high heat flux q_s. Table 1 shows the thickness, the estimated value of the thermal conductivity and the resulting specific thermal resistances (per unit area) of the corresponding layers of the materials. The total value of the specific thermal resistance is equal to $r_{B\text{-}V} \approx (6\text{-}10)\cdot 10^{-5}$ K·m²/W. With the heat flux of $q_s = 20$ W/cm², the temperature drop can be estimated as $\Delta T_{B\text{-}V} = q_s \cdot r_{B\text{-}V} \approx 12\text{-}19$ °C.

The estimation shows that the main contribution to the temperature drop $\Delta T_{B\text{-}V}$ is made by the ceramic layers under the heat source. The total thickness of these layers cannot be reduced due to technological difficulties.

The wetted wick layer in the heat pipe also makes a significant contribution to the thermal resistance. Table 1 presents a rough estimation. The transfer of the input heat flux through the wetted wick to the evaporation zone at the liquid-vapor interface is a complex process, which is determined by the geometry and properties of the porous structure, thermal properties of the working fluid and value of the heat flux. In addition, the thermal resistance of the wetted wick may decrease with increasing the input heat flux. This may occur gradually with the recession of the liquid [8] or abruptly due to a transition from the evaporation to the boiling regime [9]. When the heat load exceeds a certain critical value, the wick begins to dry out, a hot spot appears and the temperature drop on the wick increases rapidly. The critical value of the heat load depends on limitations for the heat pipe operation (capillary limit, boiling limit).

Table 1. The material layers in the area of the high heat flux.

Material layer	d (mm)	λ (W·m^{-1}·K^{-1})	r (10^{-6}×K·m^2·W^{-1})
Solder	2 × 0.1	50	4
LTCC with thermal vias	0.6	15-20	30-40
LTCC	0.05	2-4	12-25
Wick	0.25	10-20	12-25
Total			**≈ 60-100**

The temperature drop $\Delta T_{V\text{-}C}$ occurs at the low heat flux q_c. The estimation of the specific thermal resistance of the material layers is given in table 2. The estimated total value is equal to $r_{V\text{-}C} \approx (1.5\text{-}1.6)\cdot 10^{-4}$ K·m²/W. For the heat flux $q_c = 1$ W/cm², the temperature drop has a low value $\Delta T_{V\text{-}C} = q_c \cdot r_{V\text{-}C} \approx 1\text{-}2$ °C.

The temperature drop ΔT_V is determined by the thermal properties of the working fluid used, design of the vapor channels and configuration of the heat input/sink areas. For acetone as a working fluid, the temperature drop is low, $\Delta T_V < 1$ °C.

In addition, it should be noted that the temperature drop in the heat pipe depends on the quality of the heat pipe filling and it can increase in the presence of non-condensable gases or in the case of an excessive amount of the working fluid.

Table 2. The material layers in the area of the low heat flux.

Material layer	d (mm)	λ (W·m^{-1}·K^{-1})	r (10^{-6}×K·m^2·W^{-1})
Wick	0.25	10-20	12-25
Solder	0.1	50	2
Kovar	0.5	20	25
Thermal grease	0.2	1.8	110
Total			**≈ 150-170**

As a result of the estimation, it can be concluded that the total thermal resistance between the high-power heat source ($S \approx 1$ cm^2) on the module and the cooler ($S \approx 25$ cm^2) will be within $R_{S\text{-}C} \approx 0.6$-$1.1$°C/W.

During the experiments, the temperature of the top of the transistor case T_S is measured (see figure 3). Inside the transistors used, a heat-dissipating semiconductor substrate is mounted on the drain terminal attached to the bottom of the case. The drain terminal is a heat sink for the hot semiconductor substrate. The hot substrate has almost no thermal contact with the ceramic case of the transistor. Therefore, the temperature of the top of the case T_S is close to the terminal temperature T_B.

5. Results of the experiments

The experimental investigations of the ceramic circuit board test sample with the embedded heat pipe were conducted under different conditions. The objectives of the investigations were as follows: 1) evaluation of the flat heat pipe operation efficiency for the spreading of the high heat flux from the small-size heat source over the entire area of the sample; 2) measurement of the maximum power of the heat source, at which the wick in the heat pipe begins to dry out; 3) measurement of the thermal resistances between the heat source and the cooler.

For the experiments, the sample of the LTCC module was installed horizontally on the cooled surface, which was a metal plate with channels for a liquid coolant. The coolant was pumped through the channels. The temperature of the coolant was controlled by means of a thermostat. The measurements were carried out for two variants of heat removal. For the first variant, only a half of the bottom surface of the sample was pressed against the cooled surface (figure 4a). In this case, the heaters used were located on the edge opposite to the cooling area (transistors 1 and 2). Such a configuration makes it possible to estimate the efficiency of the heat pipe by the distribution of the heat flux over the entire area of the sample. For the second variant, the sample was installed by the entire bottom surface of the sample on the cooler (figure 4b). The second configuration allows determining the values of thermal resistances under the best heat removal conditions. For brevity, both of the heat removal configurations are further referred to as "half area cooling" and "entire area cooling".

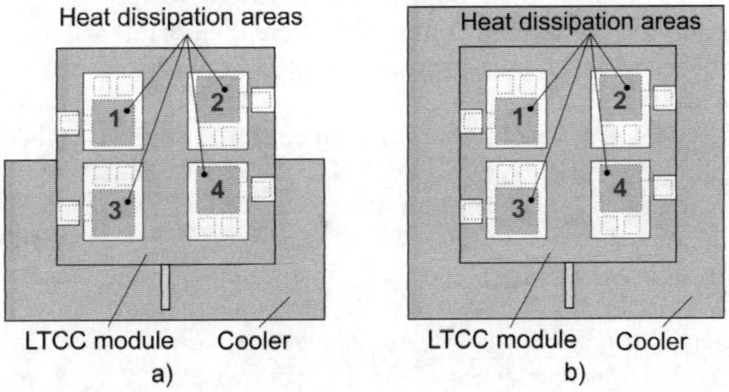

Figure 4. Variants of the heat removal configurations from the sample in the experiments: a) heat removal from the half bottom surface of the sample (half area cooling); b) heat removal from the entire bottom surface of the sample (entire area cooling).

Two series of experiments were carried out. The first series of experiments was carried out using an infrared camera. Thermal images allow estimating the performance of the flat heat pipe for the heat flux spreading from the transistors over the entire area of the sample. The experiments were performed under atmospheric conditions at the cooler temperature of 20 °C.

The second series of experiments was carried out in vacuum. The purpose of the experiments was to investigate the influence of the cooler temperature on the efficiency of the heat pipe operation and on the values of the thermal resistances. The experiments in vacuum allow one to exclude the heat exchange with the ambient air, and also to conduct studies at temperatures below zero. During the tests in vacuum, the infrared camera was not used, with the temperatures being measured by thermocouples (T-type).

During the experiments for a given cooler temperature, the power of one or two heat sources was increased stepwise. For each power value after reaching a stationary state, the temperatures of the transistors T_{S1}-T_{S4} (on the top of case) and the surface temperature of the cooler T_C were measured. When using the infrared camera, the temperature of the transistor was determined as the maximum value for the region of the transistor case on the thermal image. In the experiments without the infrared camera (in vacuum), thermocouples were placed on the transistor cases at the hottest points (determined by the thermal images obtained previously).

To analyze the measurement results, the temperature T_V was also used, which was calculated as the average temperature of the non-used transistors (without heat dissipation). The value of T_V was considered as the value of the vapor temperature in the heat pipe. The following is an estimate of the deviation of the measured non-used transistor temperature from the vapor temperature. This deviation is determined by the heat leakage Q_{TA} from the non-used transistor to the environment and thermal resistance R_T between the temperature measuring point and vapor: $\Delta T_{TV} = Q_{TA}/R_T$.

The value of $R_T \leq 5$ °C/W was estimated from the internal structure of the transistor and thermal resistance of the layers under the transistor (table 1). The value of Q_{TA} depends on the experimental conditions. For atmospheric experiments (using the infrared camera), the Q_{TA} value is determined by the convective heat exchange with the ambient air and radiation between the transistor and the environment. For the transistor with the temperature of 40 °C and the ambient air with the temperature of 20 °C, the Q_{TA} value does not exceed 0.05 W. For the experiments in vacuum, the thermocouples were pressed to the transistor using a heat-insulating material.

The assessment of the possible heat leakage Q_{TA} through the clamping structure gives a range from minus 0.15 W (heat flow to the transistor at minus 40 °C) to 0.05 W (at +40 °C). Thus, the deviation ΔT_{TV} ranges from minus 0.3 °C to +0.8 °C, depending on the temperature of the non-used transistor. In addition, it can be noted that in the experiments the measured temperatures of the non-used transistors were close to each other within 1-2 °C (except for the experiments at $T_C = -40$ °C when the temperature difference reached up to 3.5 °C).

Based on the results of the temperature measurements, the following thermal resistances were calculated: the total thermal resistance between the heat-dissipating transistor and the cooler $R_{S\text{-}C} = (T_S - T_C) / Q_S$, thermal resistances between the transistor and the vapor $R_{S\text{-}V} = (T_S - T_V) / Q_S$ and between the vapor and the cooler $R_{V\text{-}C} = (T_V - T_C) / Q_S$. The power Q_S was calculated from the electric current and voltage drop between the drain and the source terminal of the transistor.

5.1. Results of the experiments in atmosphere using infrared camera

Figure 5 presents the measured dependences of the thermal resistances vs. power of a single transistor (transistor 1 in figure 4). The blue and red lines correspond to various configurations of the heat removal from the module: entire area cooling and half area cooling, respectively (labeled "entire" and "half").

At the heat loads below 15-20 W, the measured thermal resistances decrease with the increasing power. The causes for this decrease can be the following: 1) the peculiarities of the heat pipe operation; and 2) the influence of the parasitic inflow heat to the module from the environment during the experiment. With the further increase in the power, the thermal resistances remain almost constant until the heat load exceeds a certain critical value, after which the thermal resistance between the transistor and the cooler $R_{S\text{-}C}$ begins to grow. The growth of the thermal resistance is connected with the wick dryout in the heat pipe. At half area cooling (red lines in figure 5), the wick dryout begins at the power $Q_{dry} = 35\text{-}40$ W. At entire area cooling (blue lines in figure 5), the power of the wick dryout exceeds the value of $Q_{dry} > 40$ W (at $Q = 45$ W, there is a slight increase in the thermal resistance value).

The optimum (minimum) value of the total thermal resistance $R_{S\text{-}C}$ is about 0.60-0.65 °C/W and 0.70-0.75 °C/W for the entire and for the half area cooling configurations, respectively. The difference is due to different areas of the heat sink from the module and therefore, different thermal resistance $R_{V\text{-}C}$ between the vapor in the heat pipe and the cooler. The value of $R_{V\text{-}C}$ is 0.1 °C/W and 0.2 °C/W for the entire and for the half area cooling configurations, respectively.

Figure 6 presents an infrared image of the module at the single transistor heat dissipation of 30 W at the cooler temperature Tc = +20 °C. The half area cooling configuration is used. The maximum temperature on the transistor case is 41 °C. The temperature of the module surface (beyond the heated transistor) is almost isothermal within 25-26 °C. This means that the heat flow from the heat source is transferred to the vapor in the heat pipe and effectively spreads to the entire area of the module.

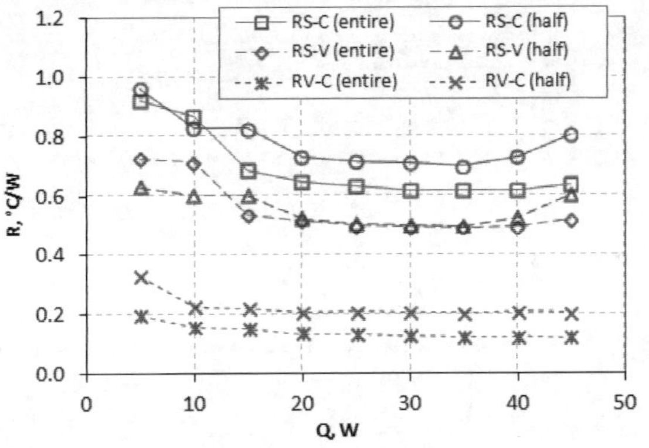

Figure 5. Thermal resistances vs. power of a single heated transistor.

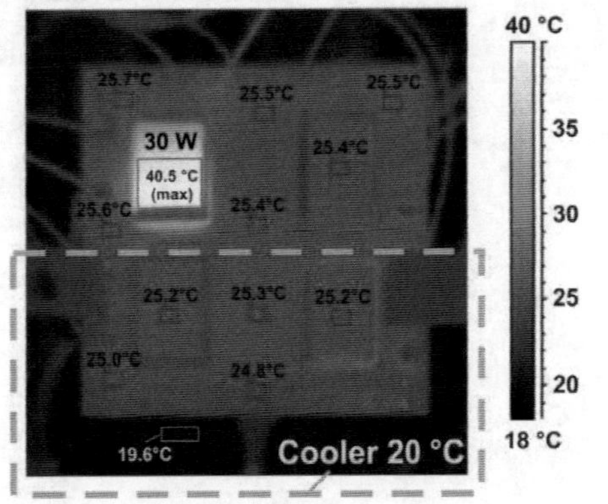

Figure 6. Infrared image of the LTCC module with the heat dissipation of 30 W.

Figure 7 shows the measured values of the thermal resistances vs. power upon simultaneously heating two transistors. The indices "1" and "2" in the notations in figure 7 correspond to transistors 1 and 2, respectively, as is shown in figure 4. The powers of both transistors in the experiment are set to be the same, $Q_1 = Q_2$. The values of $R_{S1\text{-}C}$, $R_{S1\text{-}V}$, $R_{S2\text{-}C}$ and $R_{S2\text{-}V}$ are calculated from the power value of the corresponding transistor Q_1 or Q_2; the value of $R_{V\text{-}C}$ is calculated from the total value of the power $Q_1 + Q_2$.

At entire area cooling (blue lines in figure 7), the values of the thermal resistances $R_{S1\text{-}C}$ and $R_{S2\text{-}C}$ are close to each other. At the power higher than 15 W per transistor, the thermal resistances are approximately the same and equal to 0.7-0.8 °C/W.

At the power of $Q = 45$ W per transistor (total 90 W), the thermal resistance for the second transistor $R_{S2\text{-}C}$ (solid square marks in figure 7) begins to grow. At half area cooling (red lines in figure 7), the thermal resistances for both transistors $R_{S1\text{-}C}$ and $R_{S2\text{-}C}$ are almost the same below the power of $Q \approx 25$ W (per each). With the further power growth, the value of $R_{S2\text{-}C}$ begins to increase rapidly. The value of $R_{S1\text{-}C}$ also begins to increase, but at a higher power, about 30 W. This is due to the fact that the heat dissipation area of transistor 2 (drain of the transistor) is farther from the heat removal area (see, figure 4a).

Figures 8 and 9 present the infrared images of the module for two simultaneously heated transistors at the half area cooling configuration. The infrared image at the transistor power of 20 W per each (total 40 W) is shown in figure 8. The temperatures of the transistor cases do not exceed 38 °C. The thermal resistances between the transistors and the condenser are equal to 0.9 °C/W. The surface temperature of the board, including the unheated transistors (number 3 and 4), has the uniform distribution within 27.0-28.5 °C. This means that the flat heat pipe effectively spreads the heat flux to the entire area of the module. The thermal resistance between the vapor in the heat pipe and the condenser $R_{V\text{-}C}$ is equal to 0.2 °C/W.

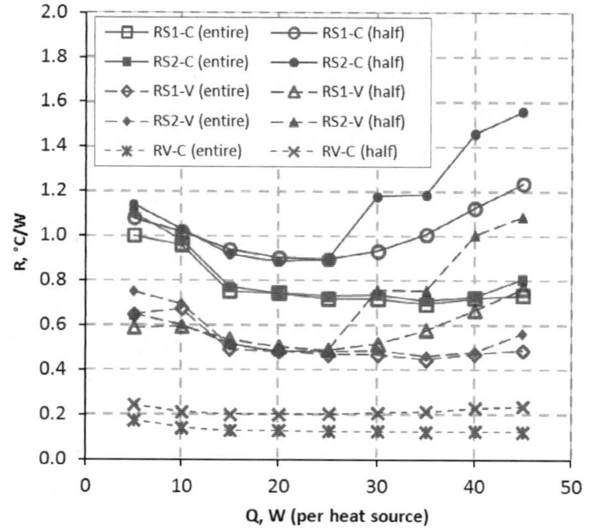

Figure 7. Thermal resistances vs. power at the heat dissipation of two transistors

Figure 9 presents the infrared images for the case of the wick dryout in the heat pipe at the high heat dissipation of 45 W per transistor (total 90 W). Hot areas can be seen around the heated transistors especially near transistor 2. This overheating occurs as a result of the wick dryout in the flat heat pipe. The temperatures of the transistors reach T_{S1} = 75 °C and T_{S2} = 89 °C. The temperature of the adiabatic section of the flat heat pipe (the vapor temperature) is in the range of approximately $T_V \approx$ 39-44 °C, which corresponds to the thermal resistance $R_{V\text{-}C} \approx$ 0.2-0.3 °C/W. In addition, it can be noted that the area with the lower temperature is observed near the filling tube. The liquid is displaced from the zone of the wick dryout. Therefore, the excessive amount of the liquid accumulates in the condensation area of the heat pipe and blocks a part of its area near the filling tube.

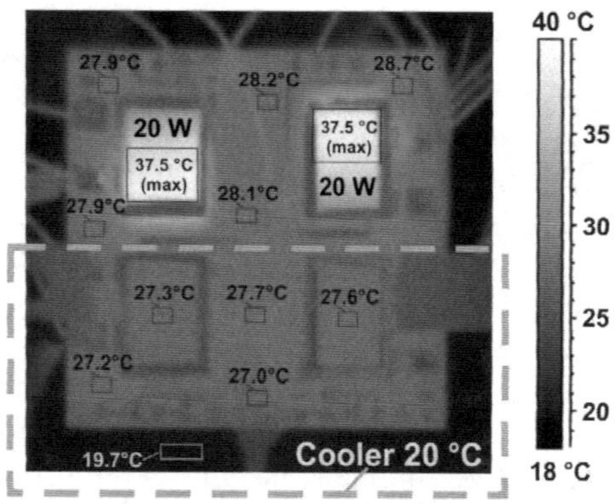

Figure 8. Infrared image of the LTCC module with the heat dissipation power of two transistors 20 W per each.

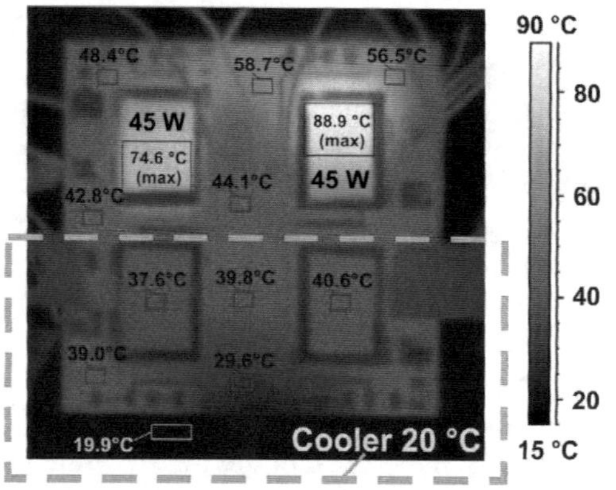

Figure 9. Infrared image of the LTCC module with the heat dissipation power of two transistors 45 W per each.

The experiments conducted in the atmosphere using the infrared imager show that the heat pipe efficiently spreads the heat flow from the transistors with the heat dissipation area of 1 cm^2 to the entire module area of 25 cm^2. This makes it possible to reduce the heat flux density and to simplify further heat removal from the module. The board outside the heated transistors has a uniform temperature close to the temperature of the heat-removing base. The heat load, at which the wick dryout begins, exceeds Q_{dry} > 40 W for the single heated transistor. With the two heated transistors the value of Q_{dry} is about 40 W per each (total 90 W) at entire area cooling and 25-30 W per each (total 50-60 W) at half area cooling.

5.2. Results of the experiments in vacuum at various cooler temperatures

Figures 10 and 11 present the measured dependences of the thermal resistances vs. heat dissipation power of the single transistor in vacuum at the entire and half area cooling configurations, respectively. The measurements were made for different temperatures of the cooler T_C: -40 ° C, -20 °C, 0 ° C and +20 ° C.

The dependences are typical. At the low heat loads, the total thermal resistance between the transistor and cooler $R_{S\text{-}C}$ decreases with the increasing heat load. After exceeding a certain value of the heat source power, the thermal resistance begins to grow. The increase in the thermal resistance is usually connected with the wick dryout due to reaching one of the limitations of the heat pipe operation.

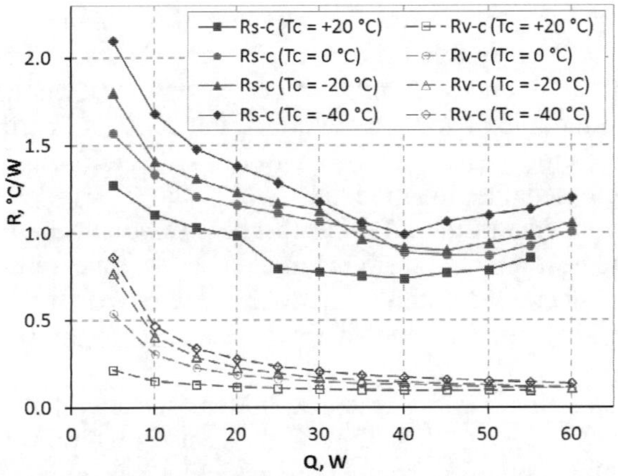

Figure 10. Thermal resistances vs. power at heat dissipation of the single transistor; heat removal from the entire bottom surface of the sample.

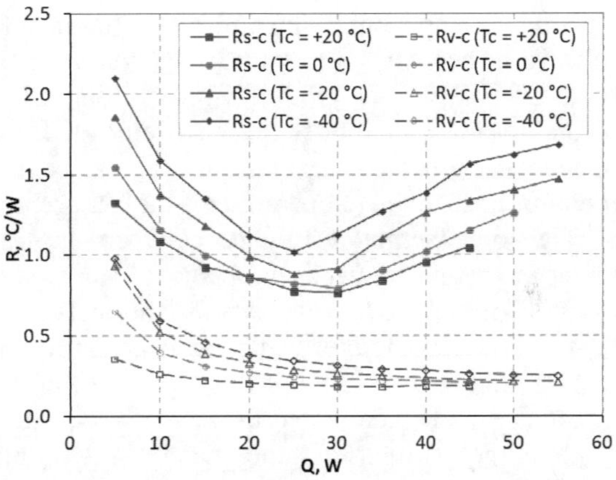

Figure 11. Thermal resistances vs. power at heat dissipation of the single transistor; heat removal from the half bottom surface of the sample.

At entire area cooling (figure 10), the thermal resistance $R_{S\text{-}C}$ begins to grow when the power exceeds the value of approximately $Q_{dry} \approx$ 35-45 W. Since the growth is slow, the beginning of the wick dryout is difficult to identify. At half area cooling (figure 11), the growth of $R_{S\text{-}C}$ is more rapid. The wick dryout occurs at the power $Q_{dry} \approx$ 30 W for T_C = +20 °C. At lower temperatures of the cooler, the value of Q_{dry} is slightly lower: at T_C = –40 °C, the $Q_{dry} \approx$ 25 W.

The values of the thermal resistance $R_{S\text{-}C}$ increase with the decreasing temperature of the cooler. At entire area cooling (figure 10) at T_C = +20 °C, the value of $R_{S\text{-}C}$ in the power range from 25 to 45 W remains almost constant and has the value of about 0.75 °C/W. At half area cooling (figure 11), at T_C = +20 °C, the thermal resistance $R_{S\text{-}C}$ has the minimum value of 0.8 °C/ W at the power of 25-30 W. When the cooler temperature is decreased to T_C = –40 °C, the value of $R_{S\text{-}C}$ increases by 30-50%.

The thermal resistance between the vapor in the heat pipe and the cooler $R_{V\text{-}C}$ (dashed lines) increases with decreasing the cooler temperature, especially at low heat loads. This is mainly due to the change in the thermal properties of acetone with the decreasing temperature. Reducing the density and pressure of the saturated vapor leads to the increase in the temperature drop along the vapor moving path in the heat pipe.

The experiments conducted in vacuum show that the heat load of the wick dryout beginning at Q_{dry} weakly depends on the cooler temperature and has the values of 35-45 W and 25-30 W at entire and half area cooling, respectively. If it is necessary to operate the module at low temperatures of the cooler, it does not require imposing additional restrictions on the heat source power.

The values of the thermal resistance $R_{S\text{-}C}$ increase with the decreasing temperature of the cooler. The causes for this increase are: 1) change in the characteristics of acetone with temperature; 2) possible presence of non-condensable gases in the heat pipe; 3) parasitic heat inflows during the experiment, especially at low operating temperatures and low heat loads. The increase in the total thermal resistance with the decreasing temperature of the cooler does not lead to an increase in the absolute value of the transistor temperature. The temperature of the heated transistor does not rise above 55 °C for the cooler temperatures from –40 °C to +20 °C at powers of 45 W and 30 W for the entire and half area cooling configurations, respectively.

6. Conclusions

A test sample of the LTCC-module with an embedded flat heat pipe is developed and investigated. The sample is a self-contained unit, which includes MOSFET transistors as miniature high heat flux electronic components. The embedded flat heat pipe provides the effective spreading of the high heat flux from the heat sources of 30-40 W power with the heat dissipation area of ≈1 cm^2 to the entire bottom surface of the module with the dimensions of 5 × 5 cm^2. A significant decrease in the density of the heat flux facilitates the heat removal from the module. The temperature of the ceramic board outside the areas of the high-power heat sources location has an almost uniform distribution and remains close to the temperature of the heat-removing base under the module.

The total thermal resistance between the transistors in the module and the heat-removing base under the module depends on the heat load, area of the heat removal from the module, and temperature of the cooler. At low heat loads the thermal resistance decreases with the increase in the power of the heat source. When the heat load exceeds a certain critical value the thermal resistance begins to increase due to drying out of the wick in the heat pipe.

At the cooler temperature of +20 °C, at the heat removal from the entire bottom surface of the module, the thermal resistance reaches the minimum value of 0.7-0.8 °C/W at the heat source power of 35-40 W. At the heat removal from the half area of the bottom surface of the module, the minimum thermal resistance of 0.8-0.9 °C/W is observed at the heat source power of 25-30 W.

The total thermal resistance increases with a decrease in the temperature of the cooler. At the cooler temperature minus 40 °C, the thermal resistance increases by 30-50% as compared to the cooler temperature of +20 °C.

The main contribution (50-70 %) to the temperature drop between the heat sources and the cooler is made by the ceramic layers under the heat sources. Though these layers contain thermal via arrays, their thermal resistance reaches 0.4-0.6 °C/W. The total thickness of these layers could not be reduced due to technological difficulties. This is the drawback of the design, which makes it difficult to transfer heat from the heat sources to the heat pipe. In the future, it is planned to improve the design, allowing one to mount electronic components through holes in the ceramic board directly to the wall of the heat pipe through a separate thin heat-conducting dielectric substrate.

The thermal resistance of the wick layer at the heat input area is the second most important point. Based on the results of the experiments, the value is estimated to be 0.1-0.3 °C/W. In the future, to reduce the total thermal resistance, it is planned to decrease the thickness of the porous layer in the capillary structure at the heat input side.

The thermal resistance between the vapor in the heat pipe and the heat-removing base under the LTCC-module has a low value of 0.1 °C/W (when removing heat from the entire bottom surface of the module). Since the flat heat pipe provides the heat spreading to the entire area and significant decrease in the heat flux, the LTCC-module can be mounted on the heat-removing base with a thermal grease or adhesive having a low thermal conductivity value.

7. References

[1] Imanaka Y 2005 *Multilayered Low Temperature Cofired Ceramics (LTCC) Technology* (New York: Springer).

[2] Bechtold F 2012 Innovation steps towards a novel and cost efficient LTCC packaging technology for high end applications *J. of Microelectronics, Electronic Components and Materials* **42** (4) pp 211–224.

[3] Zampino M A 2001 Embedded heat pipes in cofired ceramic substrates for enhanced thermal management of electronics *Ph.D. Florida International University*.

[4] Jones W K, Liu Y and Gao M 2003 Micro Heat pipes in Low Temperature Cofire Ceramic (LTCC) Substrates *IEEE Transactions on Components and Packaging Technologies* **26** (1) pp 110–115.

[5] Suntsov S B, Derevyanko V A 2016 Multilayer ceramic plate cooling device, *Russian Patent* 2605432.

[6] Faghri A 1995 *Heat pipe science and technology* (London: Taylor & Francis).

[7] Zampino M A, Kandukuri R and Jones W K 2002 High performance thermal vias in LTCC substrates *Eight Intersociety Conf. on Thermal and Thermomechanical Phenomenon In Electronic Systems* pp 179–185.

[8] Wong S C, Liou J H and Chang C W 2010 Evaporation resistance measurement with visualization for sintered copper-powder evaporator in operating flat-plate heat pipes *Int. J. Heat Mass Transfer* **53** pp 3792–98.

[9] Weibel J A, Garimella S V and North M T 2010 Characterization of Evaporation and Boiling from Sintered Powder Wicks Fed by Capillary Action *Int. J. of Heat and Mass Transfer* **53** pp 4204–15.

IOP Conf. Series: Materials Science and Engineering 1139 (2021) 012003 doi:10.1088/1757-899X/1139/1/012003

Water-filled heat pipes for CubeSat thermal control

H J van Gerner[1*], H Brouwer[2], Z de Groot[2] and J Guo[3]

[1]NLR – Royal Netherlands Aerospace Center, Voorsterweg 31 Marknesse, Netherlands
[2]ISIS – Innovative Solutions In Space, Motorenweg 23 Delft, Netherlands
[3]Faculty of Space Engineering, University of Technology Delft, Netherlands

Abstract. Currently, the amount of electrical power that is available for CubeSat's is very small and for this reason, simple thermal conductance through the frame of the CubeSat is sufficient for most CubeSat missions. However, deployable solar panels have been developed recently and peak powers up to 40W can now be generated. This higher generated electrical power results in more waste heat and potentially too high temperatures inside the CubeSat. For this reason, the use of water-filled heat pipes is studied, since these are cheap, widely commercially available, and can be bent in the desired shape. Both the condenser and evaporator thermal resistance and the total heat transfer capacity of these heat pipes have been measured for a wide range of temperatures with a unique automated setup that uses Peltier elements to control the temperature. Furthermore, the heat pipes have been subjected to multiple freeze/thaw cycles and start-ups from a frozen state. After these successful tests, a heat pipe was integrated in a CubeSat and tests were carried out in several orientations. The tests show that commercially available water-filled heat pipes are suitable for CubeSat thermal control.

1. Introduction

A CubeSat is small satellite which is made from a number of 10 cm x 10 cm x 10 cm units. Currently, the amount of electrical power that is available for CubeSat's is very small (~7W for a 3-unit CubeSat) and for this reason, simple thermal conductance through the frame of the CubeSat is sufficient for most CubeSat missions. However, deployable solar panels have been developed recently and peak powers up to 40W can now be generated [1]. This higher generated electrical power results in more waste heat and potentially too high temperatures inside the CubeSat. Ammonia-filled aluminum heat pipes are commonly used for larger satellites [2]. However, these are not suitable for CubeSat's because of their high costs. For this reason, the use of water-filled copper heat pipes is studied, since these are cheap, commercially available, and can easily be bent in the desired shape. Tests were carried out on axial-grooved and sintered heat pipes on the thermal resistance and heat transfer capacity of these heat pipes. These test results were used for the thermal model of a CubeSat with heat pipe [3]. The results discussed in this paper are for heat pipes with a length of 200 mm and an outer diameter of 6 mm. These heat pipes were bought from the German company Quick-Cool. Several heat pipes were cut in order to determine heat pipe characteristics and to study possible damage due to

* Corresponding author: Henk.Jan.van.Gerner@nlr.nl

freezing cycles. Figure 1 and Figure 2 shows cross-sections of two heat pipes. The work described in this paper has been carried out by Hugo Brouwer as part of his MSc. thesis [3].

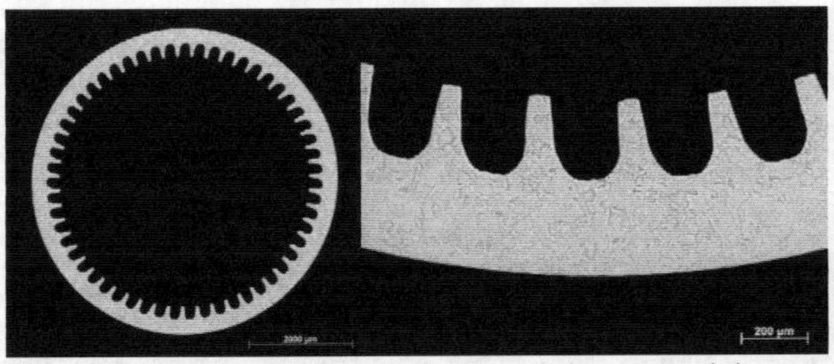

Figure 1. Cross-sections of an axial grooved heat pipe.

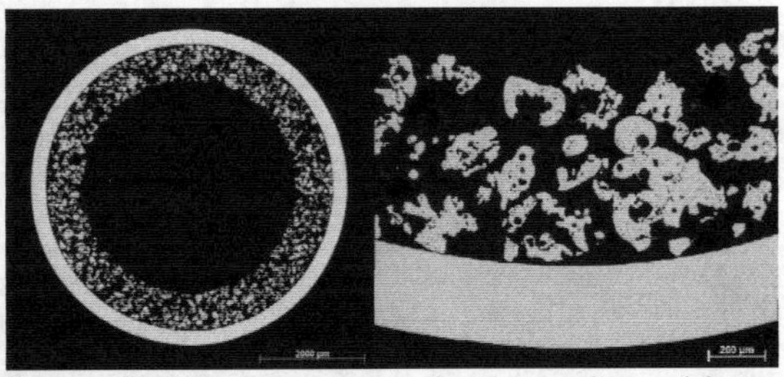

Figure 2. Cross-sections of a heat pipe with a sintered wick.

2. Test setup

Figure 3 shows the schematic drawing of the setup that is used to test the heat pipes, and Figure 4 shows a CAD drawing. The heater section consists of two copper blocks with cartridge heaters. The heat sink consists of two copper blocks between two Peltier elements which are cooled with a thermostat bath. Between the evaporator and condenser section is an adiabatic section with a length of 140 mm. The Peltier elements are used to maintain the saturation temperature of the heat pipe at the desired value, independent of the heat input (in contrast to more common setups, where the heat pipe temperature varies with the applied heater power). Two thermocouples are used to measure the evaporator section and condenser section temperature. These thermocouples have a diameter of 0.5 mm, and are inserted holes with a dimeter of 0.6 mm. The adiabatic section has three thermocouples attached to the outside of the heat pipe. Under most conditions, the inside wall of the adiabatic section of the heat pipe is wetted with liquid water (with a strong thermal coupling to the vapor inside the heat pipe) and these three temperature sensors on the adiabatic section have approximately the same temperature, which is equal to the saturation temperature inside the heat pipe. When the inside of the heat pipe is not wetted with liquid water (e.g. due to a puncture, freezing, or dry-out), a large thermal gradient between the three temperature sensors can occur due to thermal conduction along the copper casing of the heat pipe. These three temperature sensors on the adiabatic section therefor give useful information about the functioning of a heat pipe.

The Peltier controller has its own dedicated temperature sensor in the middle of the heat pipe casing. The system is fully automated, such that a large dataset can be obtained with relative small effort. Figure 5 shows a photo of the setup with the insulation partly removed. In order to measure the heat leak from the setup. a test was carried out with a disabled heat pipe (with a hole drilled in the heat pipe casing). The following heat leaks were observed:

Table 1. Measured heat leak.

Heater power	Evaporator temperature
-	**20°C (ambient temperature)**
0.7 W	**40°C**
1.5 W	**60°C**
2.3 W	**80°C**

The heat leak scales linearly with the temperature difference between the heater block and the ambient temperature and is approximately 0.0375 W/K: This heat leak is negligible compared to the thermal conductance of a functioning heat pipe. The thermal conductance of the copper casing over

140 mm can be calculated to be 0.02 W/K. Most of the tests were carried out in horizontal orientation, although some measurements were carried out in a vertical or tilted orientation to investigate the influence of gravity.

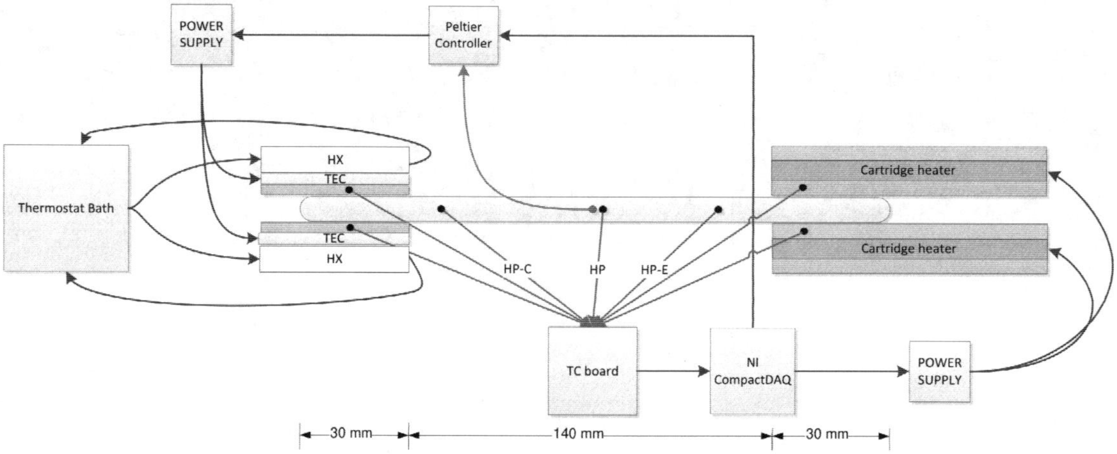

Figure 3. Schematic drawing of the test setup.

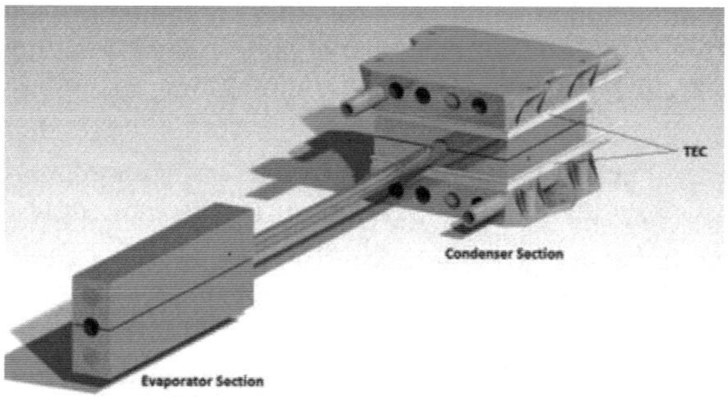

Figure 4. CAD drawing of the test setup.

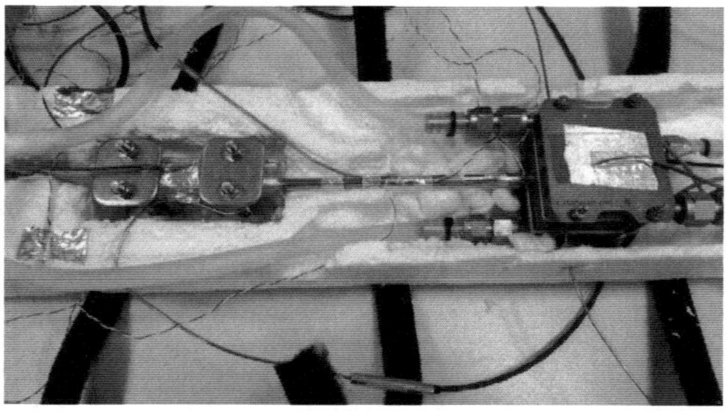

Figure 5. Photo of the test setup with insulation partly removed.

3. Capillary limit

There are several physical limits that restrict the heat transport capacity of heat pipes: the capillary limit, viscous limit, sonic limit, entrainment limit and the boiling limit [4, 5]. The capillary limit is often the most restrictive and can easily be calculated for axial grooved heat pipes. The maximum capillary pressure in an axial groove is

$$\Delta p_{\mathrm{cap}} = \frac{2\sigma \cos\alpha}{d} \tag{1}$$

where σ is the surface tension of water, α is the contact angle between copper and water (~27°), and d is the width of the liquid channel (0.2 mm). The wicking height of the axial grooved heat pipe can then be calculated to be 66 mm. The effective pore diameter of the sintered heat pipe is approximately 0.1 mm, which results in a wicking height of approximately 26 cm. The frictional pressure drop due to laminar liquid flow in a square channel in an axial grooved HP is derived from the Hagen–Poiseuille equation:

$$\Delta p_{\mathrm{friction}} = \frac{32\, \mu L\, P}{\rho N d^4 H_{lv}} \tag{2}$$

where μ is the dynamic viscosity, L is the liquid transport length (approximately 170 mm), P is the heat input, ρ is the liquid density, N is the number of channels (55), d is the width and height of the channels (0.2 mm) and H_{lv} is the heat of evaporation. The fluid properties are obtained from [6]. By combining equation (1) and (2), the capillary limit of a heat pipe in horizontal orientation can be calculated. This calculated capillary limit is compared to the measured values in the next section. The capillary pressure scales linearly with d, while the frictional pressure scales with d^4. This means that a smaller d results in a much smaller heat transport capacity (but a higher wicking height).

4. Thermal resistance measurements

4.1. Axial grooved heat pipe

The test setup was used to measure the heat transfer characteristics of the heat pipes. Figure 6 and figure 7 show the applied power and measured temperature during the test with the axial grooved heat pipe. The adiabatic section of the heat pipe has a nearly uniform temperature (the three temperature sensors on the heat pipe give nearly the same temperature, which is equal to the saturation temperature inside the heat pipe), which is held at a constant value by the Peltier elements, while the heat load is increased in small steps. This process is repeated for several heat pipe temperatures (e.g. 5 °C, 10 °C, 20 °C etc.). From these measured temperatures and heat input, the thermal resistance of several sections of the heat pipe can be derived. The thermal resistance of a heat pipe can be split in three parts:

1. Thermal resistance of the evaporator section i.e. between the temperature sensor in the heater (T_{evap}), and the adiabatic section ($T_{HP\text{-}E}$)
2. Thermal resistance of the adiabatic section, i.e. between the temperature sensor on the heat pipe near the heater section ($T_{HP\text{-}E}$) and near the condenser section ($T_{HP\text{-}C}$)
3. Thermal resistance of the condenser section i.e. between the adiabatic section ($T_{HP\text{-}C}$) and the condenser block (T_{cond})

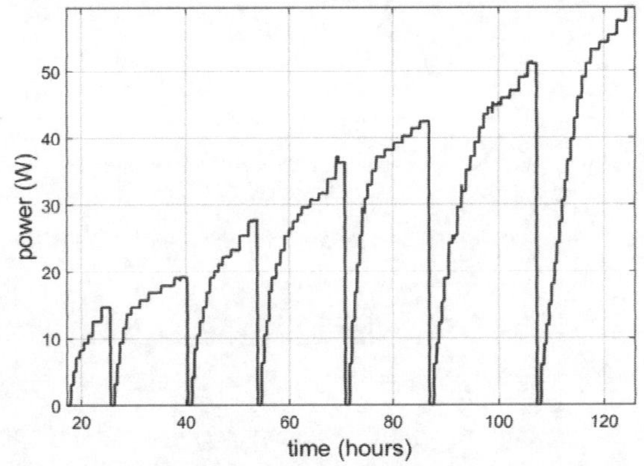

Figure 6. Applied heater power on the heat pipe.

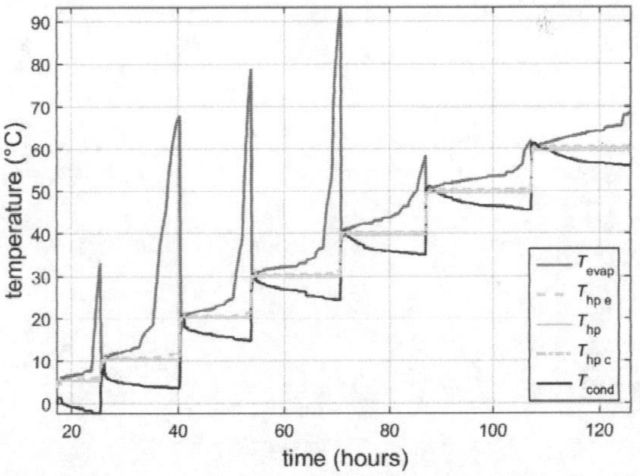

Figure 7. Measured temperatures on the copper heater block (red), three heat pipe locations (green), and the copper condenser block (blue).

The thermal resistance of the evaporator section is shown in Figure 8. The evaporator resistance decreases with temperature down to a value of 0.06 K/W. The resistance is relatively independent of heat input, until the capillary limit of the heat pipe is reached which results in a sharp increase in the resistance. For higher heat pipes temperatures (40° C, 50 °C and 60 °C) there is a good agreement between the calculated capillary limit and the measured sharp increase in thermal resistance. However, at lower heat pipe temperatures, the increase in thermal resistance occurs at a lower heat input than the calculated capillary limit.

Figure 9 shows the thermal resistance of the adiabatic section. The thermal resistance is as low as 0.02 K/W. As a result of this low thermal resistance, the temperature difference over the adiabatic section is less than 0.5 °C for most measurements (see Figure 7). The thermal resistance of the copper heat pipe casing can be calculated to be 46 K/W, i.e. a functioning heat pipe 'conducts' heat more than 2000 times better than an empty heat pipe.

Figure 10 shows the thermal resistance of the condenser section. The condenser resistance is very high for low temperatures, so at low temperatures, the total resistance of a heat pipe is dominated by the condenser section. This can also be observed in Figure 7. This high resistance might be caused by the presence of a small amount of non-condensable gases inside the heat pipe which can block part of the condenser at low temperatures. Figure 11 shows the total thermal resistance of an axial grooved heat pipe.

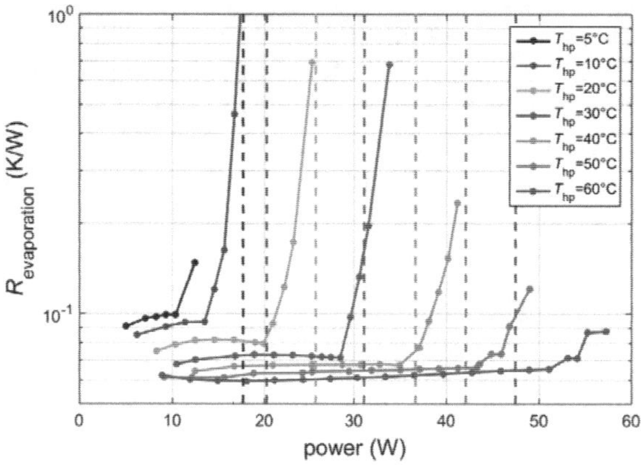

Figure 8. Condenser thermal resistance for an axial grooved heat pipe.

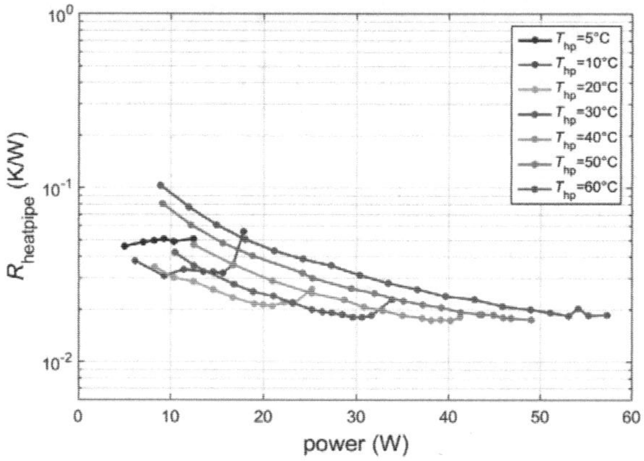

Figure 9. Total thermal resistance of an axial grooved heat pipe.

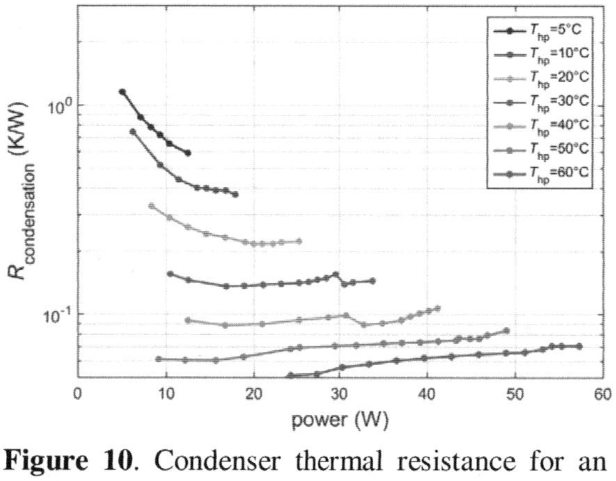

Figure 10. Condenser thermal resistance for an axial grooved heat pipe.

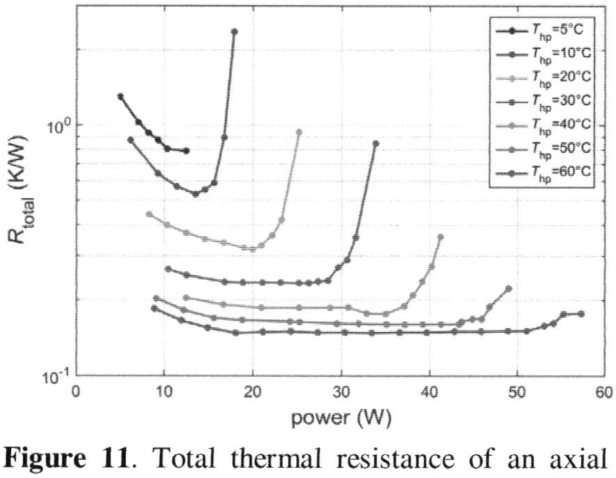

Figure 11. Total thermal resistance of an axial grooved heat pipe.

It is expected that the thermal resistance of the evaporator section and condenser section is inversely proportional to the interface area (since the thermal resistance does not strongly scales with heat flux). For heat pipe temperatures of 40°C or lower, the condenser thermal resistance is dominant over the other resistances, so the condenser interface area should be increased to lower the total resistance.

The decreasing evaporator and condenser thermal resistance at increasing temperatures results in the interesting phenomenon that the heater temperature can become lower when the heat sink temperature is increased (while maintaining the same heater power). For example, at a heater power of 36 W and a heat sink temperature of 25 °C (heat pipe temperature is 30 °C), the evaporator block rises to over 90 °C. At the same heater power but with a heat sink temperature of 57 °C (heat pipe temperature is 60 °C), the evaporator block temperature is 62.6 °C. This means that when a heat pipe is applied, the heat source temperature can be decreased by increasing the heat sink temperature.

4.2. Sintered heat pipe

In the previous section, the heat transfer characteristics of an axial grooved heat pipe have been determined. Similar measurements have been carried out for sintered heat pipes with the same outer dimensions. Figure 12 shows the measured thermal resistance in the evaporator section, Figure 13 shows the thermal resistance of the adiabatic section, Figure 14 shows the thermal resistance in the condenser section and Figure 15 shows the total thermal resistance of the sintered heat pipe. For the sintered heat pipes, the capillary limit cannot be calculated since the permeability of the sintered material is not known. Compared to the axial grooved heat pipes, the sintered heat pipes have a much lower heat transport capacity because the frictional liquid pressure drop in the sintered wick is much larger than in the axial grooves. Also, the increase in thermal resistance when the capillary limit is reached is less sharp than for an axial grooved heat pipe.

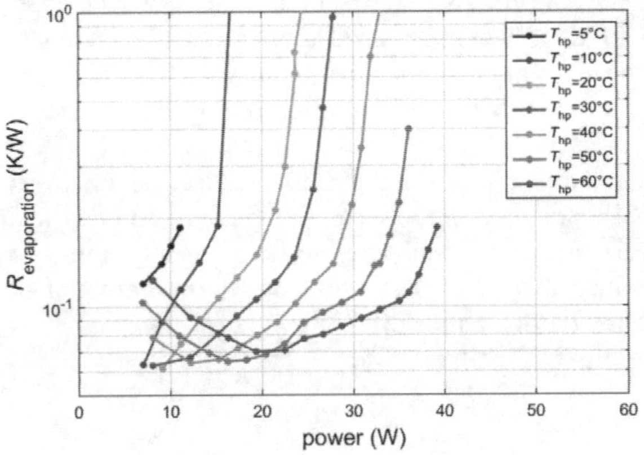

Figure 12. Evaporator thermal resistance for a sintered heat pipe.

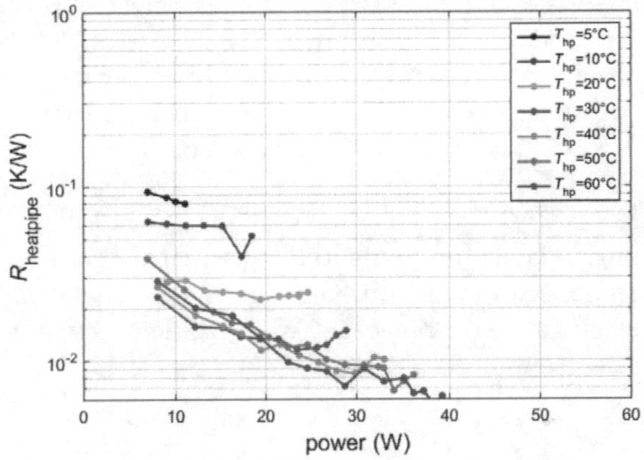

Figure 13. Thermal resistance of adiabatic section for a sintered heat pipe.

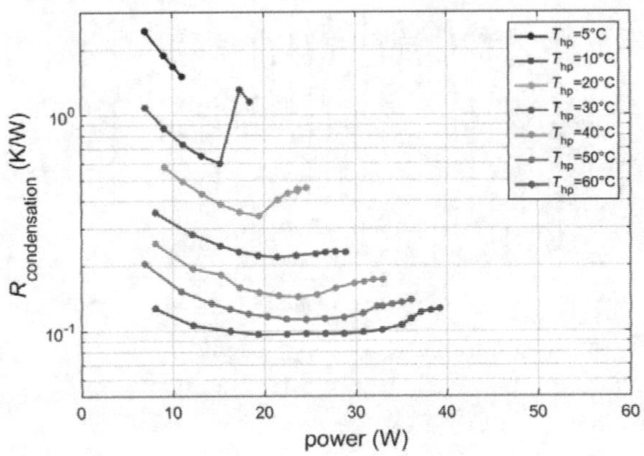

Figure 14. Condenser thermal resistance for a sintered heat pipe.

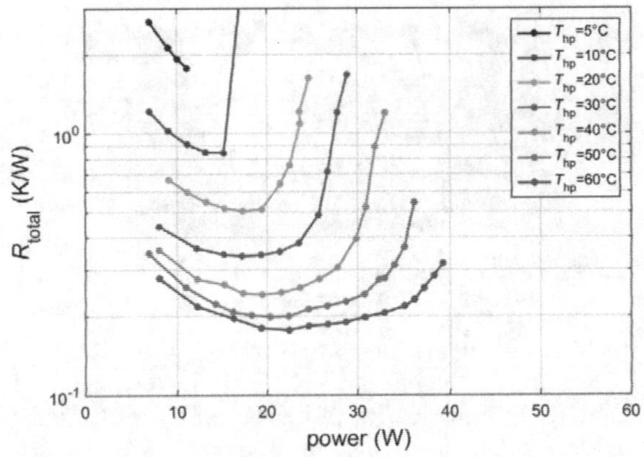

Figure 15. Total thermal resistance of a sintered heat pipe.

5. Freeze/thaw cycles

The radiator temperature of a CubeSat can become lower than 0°C, depending on orbit, radiator size and heat dissipation [3]. From literature, it is known that water-filled heat pipes can be damaged by multiple freeze-thaw cycles [7]. When the condenser section freezes, all the water in the heat pipe will accumulate in the condenser section, and when the condenser section thaws again, the heat pipe can be damaged by expansion of the liquid. Figure 11 in paper [7] shows a photo of a heat pipe that has ruptured at the condenser end of the heat pipe after 60 freeze/thaw cycles. This heat pipe has a length of 305 mm, an outer diameter of 4 mm and has a sintered wick. It was concluded that this damage was due to expansion of liquid from the frozen state. However, the heat pipes intended for CubeSats are shorter and have a larger diameter, and this makes it less likely that the heat pipe is damaged by freeze/thaw cycles.

Furthermore, an axial grooved heat pipe might be better suited to handle freeze/thaw cycles than a heat pipe with a sintered wick. For this reason, 100 freeze/thaw cycles were carried out for the axial grooved heat pipe (see Figure 16) and 82 for the sintered heat pipe (the test sequence for the sintered heat pipe was stopped before 100 cycles were reached because of a software crash). In each cycle, the condenser is set to -15 °C for 15 minutes, and then to 10 °C for another 10 minutes. Every 10 cycles, the condenser temperature is increased to 30 °C and 20 W of heater power is applied for 15 minutes to check the functioning of the heat pipe. Figure 17 shows a short section of the test with 10 cycles and two functional tests. When the condenser section is set to -15°C, the heat pipe temperature quickly drops to just above 0°C, after which the heat pipe stops functioning and the heat pipe slowly cools further due to conduction through the copper casing. Note that the evaporator section never drops below 0 °C because of the very small heat leak (~0.7 W, see table 1) between the ambient and the heat pipe. This corresponds to the calculated heat transport by conduction through the copper casing. For an application in CubeSats, this is very interesting, since it implies that only a very small amount of electrical heater power (~1W) is required to maintain the payload above 0 °C.

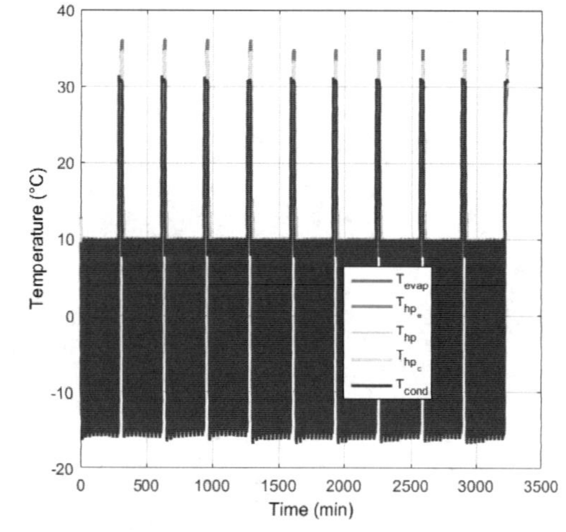

Figure 16 Measured temperatures during 100 freeze/thaw cycles. Every 10 cycles, the condenser temperature is increased to 30°C and 20W of heater power is applied for 15 minutes to check the functioning of the heat pipe.

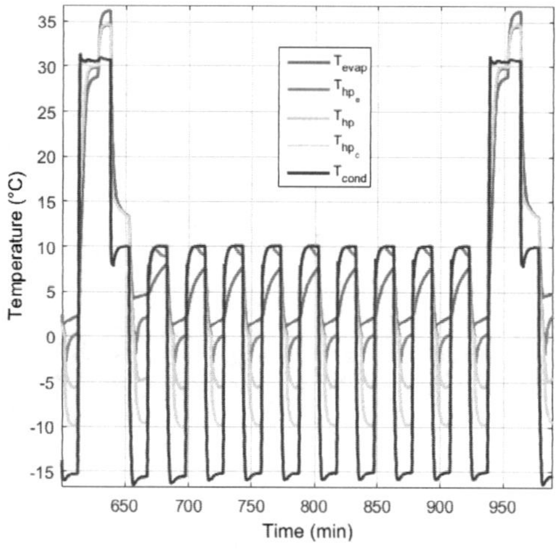

Figure 17. Zoom in on 10 cycles and 2 functional tests. See figure 3 for the location of the temperature sensors.

After the thermal cycling tests, both the axial grooved and sintered heat pipe functioned properly. Furthermore, cross section of the heat pipes (similar to Figure 1 and Figure 2) were made and the heat pipes were internally inspected for damage. From these tests, it was concluded that the freeze/thaw cycles did not damage the heat pipes.

Start-up after freezing

Another issue is start-up of the heater from a frozen state. When the condenser section of the heat pipe is frozen, all water has accumulated in the condenser section and no liquid is present in the evaporator section. When a heat load is applied (by the payload), the heat will have to be transported by conduction through the copper housing until the condenser section is thawed and the heat pipe starts functioning again. However, since the thermal resistance of a non-functioning heat pipe is very high (~46 W/K), the payload might overheat before the heat pipe starts to function. Whether a heat pipe can be started from a frozen state without local overheating depends on the thermal inertia of the payload, the applied power, and the length and thermal conductance of the heat pipe casing [8]. Figure 18 shows the measured temperatures during a start-up test of an axial-grooved heat pipe with a frozen condenser section. After t=6 minutes, 10W of heat input is applied and the temperature of the evaporator section starts to rise. There is a temperature gradient over the adiabatic section of the heat pipe, which indicates that heat transport is mainly via conduction through the copper casing. Around t=17 minutes, part of the condenser starts to thaw, and the heat pipe starts to operate. At t=26 minutes, all the sensors on the adiabatic section have (nearly) the same temperature, which indicates that the heat pipe is functioning. This test shows that a start-up from a frozen state is possible without overheating of the heater section.

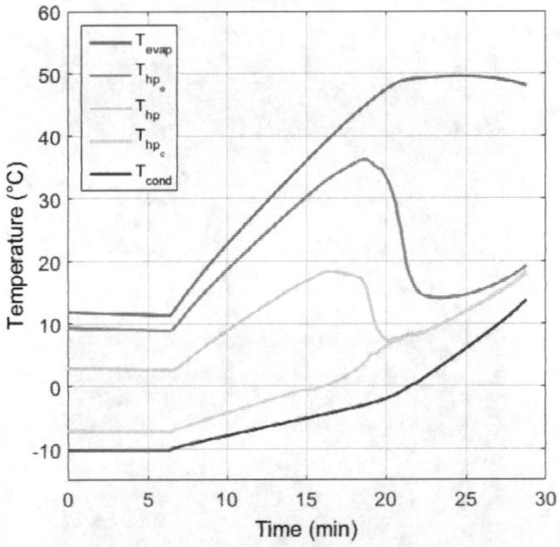

Figure 18. Measured temperatures during start-up from a frozen state.

6. Bending of heat pipes

For application of heat pipes in CubeSats, the heat pipes have to be bend. Two sintered, two axial-grooved, and two mesh heat pipes were bent (see Figure 19) with a bending tool with a bending radius of three times the heat pipe outer diameter. The heat transport capacity of these heat pipes was measured before and after the bending. No significant influence of the bending on the thermal performance of the axial grooved and sintered heat pipes was found. The heat transport capacity of the mesh heat pipe was greatly diminished after the bending. From these tests, it was concluded that the heat pipes can be bent with this bending radius.

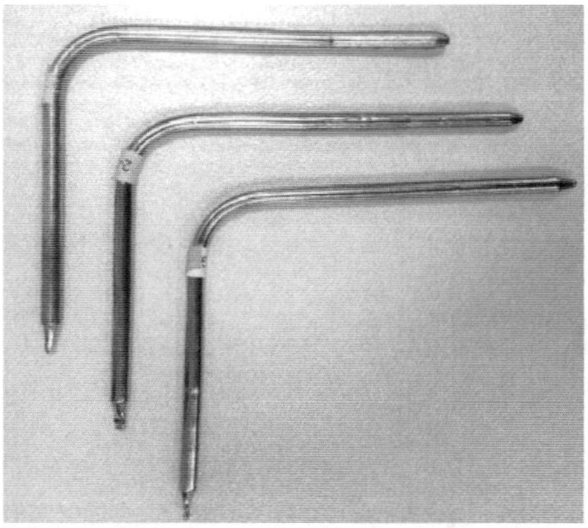

Figure 19. Photo of 3 bended heat pipes.

7. Integration in CubeSat

After the successful thermal and bending tests, a heat pipe was integrated in a CubeSat (see Figure 20). The bending tool was not available during integration and the heat pipes were bended by hand. Tests were carried out in several orientations (both gravity-assisted and anti-gravity). Furthermore, start-up tests from a frozen condition were carried out. The tests (which are described in more detail in [9]) show that commercially available water-filled heat pipes are very suitable for CubeSat thermal control.

Figure 20. Photo of a CubeSat (with some panels removed) with a heat pipe.

8. Conclusions

The thermal resistance and heat transport capacity of heat pipes was measured and the correspondence with calculated values is reasonably well. The measured values have been used in a thermal model of a CubeSat to analyse its thermal behaviour [3]. No damage to the heat pipes was observed after 100 freeze/thaw cycles. The freezing of water in the heat pipe can even be an advantage, since it can protect the payload against too low temperatures. The tests show that commercially available water-filled heat pipes are suitable for CubeSat thermal control.

9. References

[1] Weeren H, Brake M T, Hamann R, Holl G and Price S 2009 Thermal Aspects of Satellite Downscaling, *Journal of Thermophysics and Heat Transfer*, **23**, 592

[2] Gilmore D G 2002 *Spacecraft Thermal Control Handbook. Vol I: Fundamental Technologies*, 2nd ed. ,The Aerospace Corporation, California

[3] Brouwer H S B 2016 Performance Characterization of Water Heat Pipes and their Application in CubeSats, MSc. Thesis, https://repository.tudelft.nl/, Delft.

[4] Peterson G P 1994 *An Introduction Into Heat Pipes: Modelling, Testing, and Applications*, John Wiley & Sons Inc.

[5] Wits W W and Riele G J 2017 Modelling and performance of heat pipes with long evaporator sections. *Heat and Mass Transfer*, **53**(11):3341-3351

[6] Lemmon E W, Huber M L, McLinden M O 2013 NIST Standard Reference Database 23: Reference Fluid Thermodynamic and Transport Properties-REFPROP, Version 9.1, National Institute of Standards and Technology, Standard Reference Data Program, Gaithersburg.

[7] Cheung K 2005 Flight Qualification of Copper Water Heat Pipes At Naval Research Laboratory, *38th AIAA Thermophysics Conference*, Toronto (Ontario) Canada

[8] Ochterbeck J and Peterson G 1994 Current Status of Frozen Heat Pipe Startup Research, *Proc. of the Tenth International Heat Transfer Conference*, Brighton, UK, 351–356.

[9] Brouwer H, Groot Z D, Guo J, van Gerner H J 2017 Solving the Thermal Challenge in Power-Dense CubeSats with Water Heat Pipes, *Small Satellite Conference*, SSC17-VII-06

IOP Conf. Series: Materials Science and Engineering 1139 (2021) 012006 doi:10.1088/1757-899X/1139/1/012006

Modelling of a screen mesh wick heat pipe using Al_2O_3 nanofluids

B Herrera[1,2*], F Chejne[2], M B H Mantelli[3], J Mejía[2] and K Cacua[1]

[1]Faculty of Engineering, Instituto Tecnológico Metropolitano, Cll. 54 A No. 30 – 01, Medellín, Colombia

[2]Faculty of Mines, Universidad Nacional de Colombia, Av. 80 No. 65 – 223, Medellín, Colombia

[3]Faculty of Mechanical Engineering, Universidade Federal de Santa Catarina, Trindade 88040-900, Florianópolis, Brazil

Abstract. In this work, a phenomenological model that considers the interaction of nanoparticles of a Al_2O_3/water nanofluid within a cylindrical mesh wick heat pipe has been developed. The model is based on Navier-Stokes equations and it uses population balance as a method to account for the interaction of nanoparticles. The model predicted values of capillary limit with nanofluids up to 32% higher than that of the heat pipe using DI water. Those values were in agreement with the values found experimentally. Although the better performance of the capillary limit in the model was a consequence of the reduction of the pore radius which increases the capillary pressure, that was not the main cause for better capillary limit in experimental conditions at lower nanofluid concentrations. On the other hand, it was found that exists an optimum concentration of nanoparticles which decreases the effective thermal resistance of the heat pipes. In this work, that concentration was 0.5% w/w.

1. Introduction

Heat pipes are an alternative technology for making compact and high effectiveness heat exchangers. However, heat pipes are subject to operating limits, which are a series of physical phenomena that stop heat transfer. Therefore, a modification of the physical properties of the working fluid or the wick structure can help to counteract the effects of the operating limits, particularly the capillary limit.

In the mid-90s, the concept of suspensions of nanoparticles in common fluids denominated as “nanofluids” was first proposed [1]. It has been shown that nanofluids modify some properties of the base fluid, such as thermal conductivity, viscosity, wetting angle and the amount of heat that can be transferred by boiling [2–7]. On the other hand, since nanoparticles can be transported together with the base fluid within the wick in a heat pipe, it is possible that nanoparticle deposition on the surface of the wick occurs. The nanoparticle layer so formed can modify the capillarity of the wick [8].

Several theoretical studies have considered the nanoparticle layer existence on the surface of the wick of heat pipes [7,9–11]. Those works found an optimum concentration of nanoparticles to achieve a low thermal resistance of the heat pipe. The optimum concentration of nanoparticles is generally

* Corresponding author: bernardoherrera@itm.edu.co

associated to the competition between two effects: on one hand, there is an extension of the evaporation surface because of the formation of the nanoparticle layer; on the other hand, there is an increase in the pressure drop along the wick due to the increased thickness of the nanoparticle layer [10]. However, those results were obtained only considering the effect of variation in the nanofluid physical properties such as viscosity and density as a function of the concentration of nanoparticles. No interactions of the nanoparticles in nanofluid to form a porous layer was considered. In other cases, the authors increased arbitrarily the diameter of the threads in the screen mesh of the wick to account nanoparticles deposited layer [11], assuming an uniform deposition.

On the other hand, population balance has been widely used to analyze how particles can be born or destroyed inside a control volume. This is particularly interesting in systems where phenomena like aggregation occur, such as the aggregation of nanoparticles in nanofluids. Therefore, accounting for the change of the nanoparticle size and concentration is useful to model changes in thermal properties of nanofluids within a heat pipe. The problem of modeling of nanofluids agglomeration and thermal conductivity using population balance has been addressed in previous researches [12,13] and it has been concluded that population balance is useful to model the stability, aggregation and growth of nanoclusters when there are interactions between the nanoparticles.

In this work, a phenomenological model that considers the interaction of nanoparticles of Al_2O_3 in a nanofluid to form a layer of nanoparticles in the wick of a cylindrical heat pipe has been developed. The model is based on Navier-Stokes equations for mass and energy, including energy balance to account for the vaporization and condensation mass flow. The Darcy – Forchheimer equation was used to model velocity and pressure profiles in the porous wick. A coupling between the population balance equations and the hydrodynamic of the liquid in the wick was used as a method to account for the interaction of nanoparticles and how this interaction influences the formation of a layer of nanoparticles over the porous wick. Validation of the model results have been performed by comparison with experimental data obtained in a heat pipe using a Al_2O_3/water nanofluid.

2. Materials and methods

2.1. Model description

The computational domain consists of the wick and wall of a cylindrical heat pipe as shown in Figure 1, since the heat pipe has been supposed to be axisymmetric. Transient, laminar and incompressible one dimensional continuity equation was solved for the liquid in the wick:

$$\frac{\partial}{\partial t}(\varepsilon \rho_l) + \frac{\partial}{\partial z}(\rho_l v_{l,z}) = \dot{m}_{L-V} \tag{1}$$

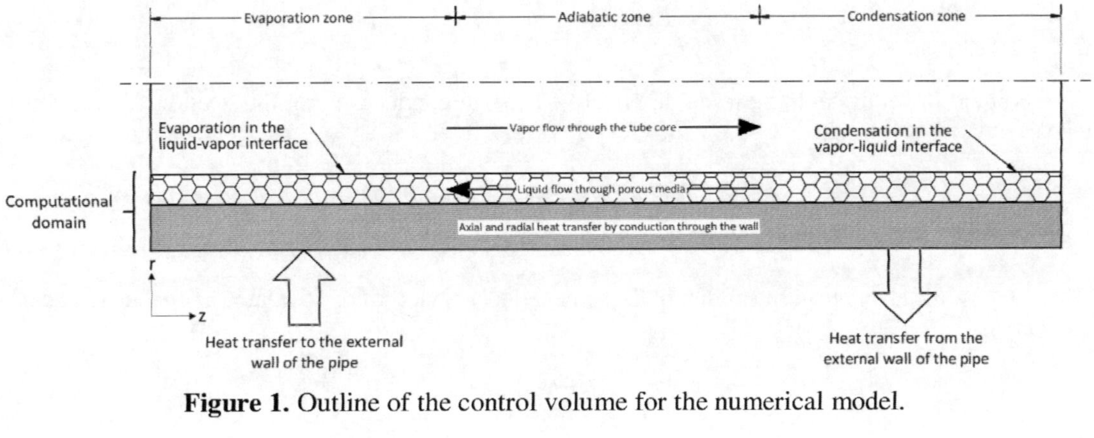

Figure 1. Outline of the control volume for the numerical model.

The Darcy – Forchheimer equation was used to calculate velocity and pressure profiles of the nanofluid within the wick:

$$\frac{dP_l}{dz} = -\frac{\mu}{K} v_{l,z} - \rho_l g \frac{dh}{dz} - \frac{C_E}{\sqrt{K}} \rho_l v_{l,z}^2 \tag{2}$$

C_E is the Ergun coefficient, which is calculated as follows:

$$C_E = \frac{1.75}{\sqrt{150\varepsilon^{3/2}}} \tag{3}$$

A 2D transient energy equation was applied to the wall around the wick:

$$\frac{\partial}{\partial t}(\rho_w C_{pw} T_w) = \frac{\partial}{\partial z}\left(k_w \frac{\partial T_w}{\partial z}\right) \tag{4}$$

In order to simplify the calculations in the central region of the heat pipe, it has been assumed that vapor is saturated at constant pressure [14]. On the other hand, an energy balance in the wick-vapor interface was performed to calculate the temperature:

$$\dot{m}_{L-V} h_{fg} + \dot{m}_{L-V} C_{pv} T_i = \frac{T_w - T_i}{R_{eff}} + \dot{m}_{L-V} C_{pl} T_i \tag{5}$$

In the wick-vapor interface, there is a constant mass flow of vapor and liquid due to the phase change of the working fluid. This flow was calculated according to the Knudsen, Hertz and Scharage equation [15]:

$$\dot{m}_{L-V} = \frac{2\eta_{ev}}{2-\eta_{ev}} \sqrt{\frac{M}{2\pi R}} \left[\frac{P_i}{\sqrt{T_i}} - \frac{P_v}{\sqrt{T_v}}\right] \tag{6}$$

The experimental accommodation coefficient η_{ev} was supposed to be 0.03 because of the evaporation inside the heat pipe occurs at low temperature [16]. Due to the phase change, it is necessary to calculate the pressure in the interface, which could be done by applying the Clausius – Clapeyron equation:

$$\frac{\bar{R}}{h_{fg}} ln\left(\frac{P_i}{P_0}\right) = \frac{1}{T_0} - \frac{1}{T_i} \tag{7}$$

As the liquid flows through the wick, an effective thermal conductivity of this section of the heat pipe can be calculated as follows:

$$k_{eff} = \frac{k_l\big(k_l + k_m - (1-\varepsilon)(k_l - k_m)\big)}{k_l + k_m + (1-\varepsilon)(k_l - k_m)} \tag{8}$$

The porosity and absolute permeability of the screened mesh wick were calculated according to equations (9) and (10):

$$\varepsilon = 1 - \frac{S_c \pi N_m d_m}{4} \tag{9}$$

$$K = \frac{d_m^2 \varepsilon^3}{122(1-\varepsilon)^2} \tag{10}$$

The boundary conditions were defined as shown in Table 1:

Table 1. Boundary conditions of the model

Boundary	Description
Evaporator wall	Constant heat flux
Adiabatic section wall	Constant heat flux = 0
Condenser wall	Convective heat transfer $\dot{Q}_{in} = h_c A_c (T_w - T_\infty)$ (11)
Upper and lowert caps wall	Constant heat flux = 0 Velocity of the liquid = 0

Thermal conductivity of the nanofluid was calculated according to heat conduction on the nanolayer [17]:

$$k_{nf} = \frac{k_{pe} + 2k_l + 2(k_{pe} - k_l)(1+\beta)^3\phi}{k_{pe} + 2k_l - (k_{pe} - k_l)(1+\beta)^3\phi} k_l \tag{12}$$

$$k_{pe} = \frac{[2(1-\alpha) + (1+\beta)^3(1+2\alpha)]\alpha}{-(1-\alpha) + (1+\beta)^3(1+2\alpha)} k_p \tag{13}$$

$$\alpha = \frac{k_{layer}}{k_p} \tag{14}$$

$$\beta = \frac{w}{r_p} \tag{15}$$

The thickness of the nanolayer w was supposed to be 2 nm [17], but a further analysis is required to find a more exact value. To complete the calculation of the thermal properties, the following equations were used to calculate viscosity [18], surface tension [19], density and specific heat [20] of the nanofluid, respectively:

$$\mu_{nf} = \mu_l(1 + 39.11\phi + 533.9\phi^2) \tag{16}$$

$$\sigma_{nf} = \sigma_l(1-\phi) \tag{17}$$

$$\rho_{nf} = (1-\phi)\rho_l + \phi\rho_p \tag{18}$$

$$C_{p,nf} = \frac{(1-\phi)\rho_l C_{p,l} + \phi\rho_p C_{p,p}}{(1-\phi)\rho_l + \phi\rho_p} \tag{19}$$

The population balance of nanoparticles was discretized according as follows:

$$\frac{\partial N_i(t)}{\partial t} = \frac{N_{i,in}}{\Delta z}\left(v_{l,z}{}^{in} - v_{l,z}{}^{out}\right) + \frac{v_{l,z}{}^{out}}{\Delta z} N_{i,r}(t) + \sum_{\substack{j,k \\ v_i \le x_j + x_k \le v_{i+1}}}^{j \ge k} \left(1 - \frac{1}{2}\delta_{j,k}\right) \beta_{jk} N_j(t) N_k(t) - N_i(t) \sum_{k=1}^{M} \beta_{jk} N_k(t) \quad (20)$$

The DLVO theory [21,22] was used to calculate the aggregation kernel (β_{jk}) to account for the birth and death of nanoparticles. In addition, absorption of nanoparticles on the surface of the wick was supposed to follow a Langmuir isothermal according to equation (21):

$$\rho_m(1-\varepsilon)\frac{\partial S}{\partial t} = \kappa_a\left(1 - \frac{S}{S_{max}}\right)c\varepsilon - \rho_m(1-\varepsilon)\kappa_d S \quad (21)$$

In this work, S_{max} was calculated according to geometrical considerations as:

$$S_{max} = \frac{\pi}{3\sqrt{3}} d_p \rho_p \rho_s \quad (22)$$

Retention of nanoparticles in the wick in every time step was calculated with the following Ordinary Differential Equations system:

$$\frac{\partial N_{i,r}(t)}{\partial t} = \kappa_a\left(1 - \frac{\varepsilon}{1-\varepsilon}\frac{m_i}{\rho_m}\frac{N_{i,r}(t)}{S_{max}}\right) N_i(t) - \kappa_d N_{i,r}(t) \quad (23)$$

The values for κ_a and κ_d were 2.3×10^{-3} and $2.0 \times 10^{-3}\ s^{-1}$, respectively [23]. Coupling of the hydrodynamic model with the population balance was as follows: the velocity and the temperature fields were calculated with equations (1) - (7) in every time step. Once those fields are known, population balance equations (20) and (23) were solved and the new distribution of nanoparticles size was calculated in every cell of the grid. As retention of nanoparticles on the wick reduces the bulk suspended nanoparticles, the porosity and pore radius of the wick were reduced in proportion to the volume of the retained nanoparticles. After modification of the later properties of the wick, capillary pressure was calculated in every cell of the domain. With the new number of suspended nanoparticles, the physical and thermal properties were calculated, and a new time step was initiated. This procedure continued until the steady state of vapor pressure in the heat pipe was reached.

2.2. *Experimental setup*

To validate the numerical results, four copper heat pipes 0.6 m long and 3/4 inches diameter were used. The length of the evaporator, adiabatic zone and condenser was 0.2 m, respectively. Inside the tubes, a copper 100 mesh/inch screen was rolled three times to form the porous wick. Heat input was provided by two electrical resistances attached to two aluminum pieces in contact with the evaporator walls. The intensity of the heating was controlled with a variable current and voltage power supply. A jacket filled with cool water at 20°C coming from a thermostatic bath was attached to condenser wall. The following number of type K thermocouples were attached in the heat pipe walls: 3 in the condenser, 3 in the adiabatic zone and 7 in the evaporator. Data of temperature profile was obtained with a data acquisition system. The inclination of heat pipe was 0°. The experimental setup can be seen in Figure 2.

CuO nanoparticles (Sigma Aldrich, < 50 nm) were suspended in DI water using an ultrasonic probe at 20 kHz for 20 minutes. The four heat pipes were filled separately with DI water and nanofluids containing 0.1%, 0.5% and 1.0% w/w of CuO nanoparticles.

After preparation of the nanofluids, a Dynamic Ligth Scatering – DLS (NanoPlus HD – Particulate Systems) analysis was conducted to each nanofluid to determine the distribution of number and size of the

IOP Conf. Series: Materials Science and Engineering 1139 (2021) 012006 doi:10.1088/1757-899X/1139/1/012006

particles and aggregates. These distributions are shown in Figure 3 after data processing using MATLAB®. This information was used later to set the initial conditions for population balance in the model.

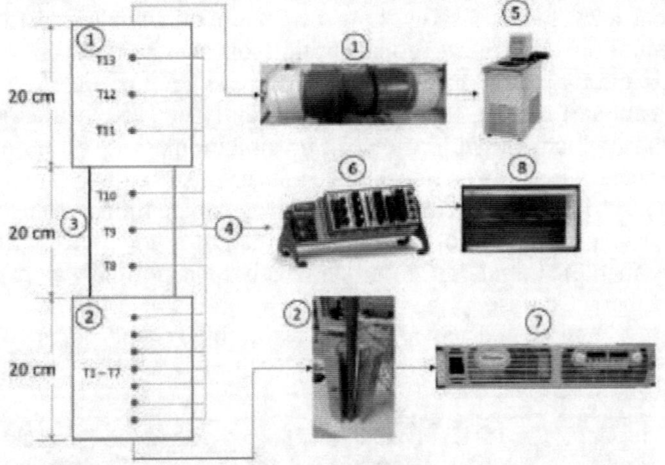

Figure 2. Test rig. 1) Condenser 2) Evaporator 3) Adiabatic 4) Thermocouple signal (T1 – T13) 5) Thermostatic bath 6) Acquisition system 7) Power supply 8) Computer

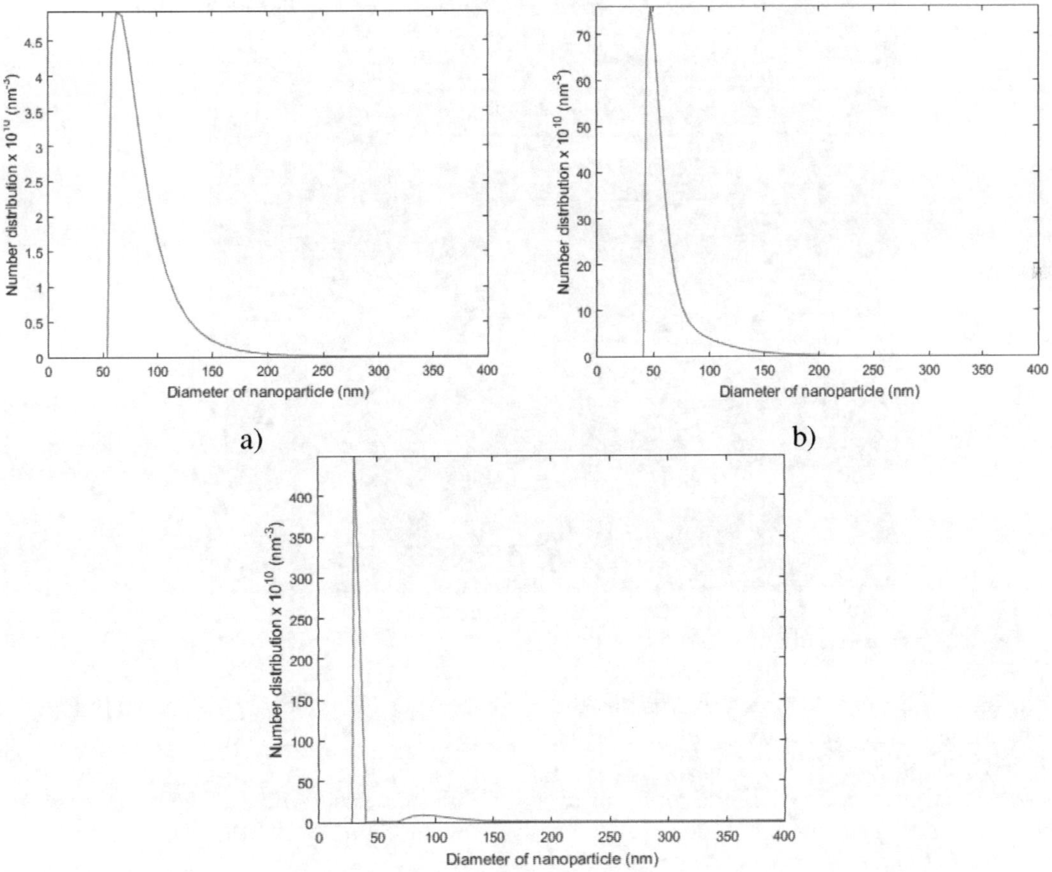

c)

Figure 3. DLS number distributions of nanoparticles in the nanofluids. a) 0.1% w/w b) 0.5% w/w c) 1.0% w/w

3. Results and discussion

There are several complications when nanofluids are used as working fluid due to the instability of the suspension of the nanoparticles. Figure 3 shows how the amount of nanoparticles suspended has a strong dependence on concentration. When the concentration of nanoparticles is low, there are more nanoparticles suspended and they form aggregates in a wide range of particle size. As concentration is higher, the number of nanoparticles suspended is consequently higher but the distribution of particle size falls in a narrow range, which is an indication of a precipitation of aggregates of higher size. Indeed, when the maximum concentration was analyzed, a secondary distribution of higher size was detected before it precipitated totally. These instabilities have incidence in the filling of the heat pipe since bigger aggregates go straight to the lower end of the tube and therefore, the concentration of nanoparticles within the wick is lower than the initial concentration which the nanofluid was prepared. In Figure 4, a SEM image of a part of the wick close to the bottom cap when the nanofluid 1.0% w/w was used. The presence of a big mass of aggregated nanoparticles which blocks the pores of the wick is mainly associated to the sedimentation of nanoparticles during the filling of the heat pipe.

Figure 4. SEM picture of the wick near the lower cap with nanofluid at 1.0% w/w

A comparison between the pressure drop and the capillary pressure in the wick is shown in Figure 5. This comparison is achieved after 2000 time steps, which are equivalent to 2000 seconds as the time step was fixed at 1 second. Every figure shows the approximate input power where capillary pressure is very close to total pressure drop in the mesh for every nanofluid concentration. If the pressure drop in the vapor is assumed as negligible, Figure 5a-d actually show the capillary limit in every case.

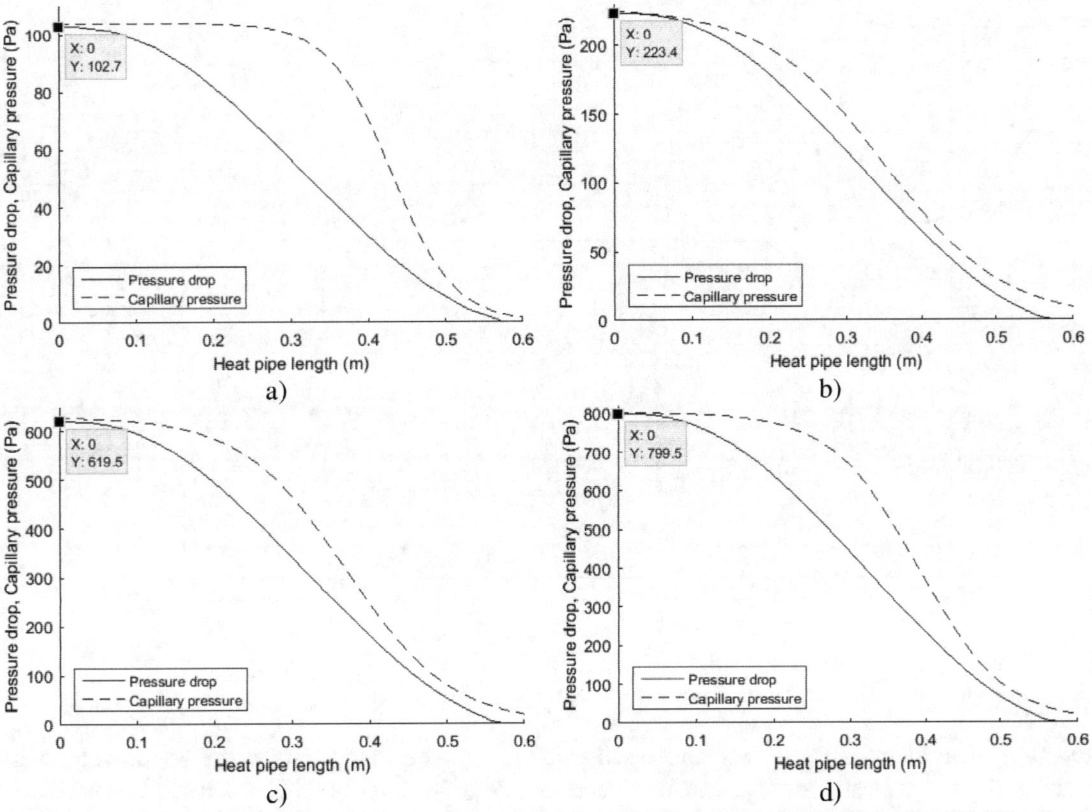

Figure 5. Pressure drop and capillary pressure in wick a) Water at 62 W b) Nanofluid 0.1% w/w at 73 W c) Nanofluid 0.5% w/w at 79 W d) Nanofluid 1.0% w/w at 82 W

Capillary limit for DI water was 62W, corresponding to a capillary pressure around 102.7 Pa. Furthermore, the capillary limit and the pressure drop increased as the concentration of nanoparticles was augmented. The numerical capillary limit was 73, 79 and 82 W for nanofluids 0.1%, 0.5% and 1.0% w/w, respectively, and the capillary pressures were 223.4, 619.5 and 799.5 Pa, respectively. This result is explained considering that the numerical model account for the attaching of nanoparticles in the mesh, which diminishes the porosity and the effective pore radius. This reduction causes an increasing of the capillary pressure as it has been established by the Young-Laplace equation:

$$P_C = \frac{2\sigma_{nf}}{r_{eff}} \tag{24}$$

Therefore, it is expected that the more nanoparticles are suspended in the nanofluid, the more of them are susceptible to deposit in the wick wires and so, the capillary limit could be increased significantly at high nanofluid concentrations. However, comparison with the experimental results shown in Figure 6 evidences that higher nanofluid concentrations actually can affect negatively the performance of the heat pipe. In order to determine the capillary limit experimentally, the effective thermal resistance of the heat pipe was calculated with the equation (25) and it was plotted against the input power in the evaporator, as shown in Figure 6:

$$R_{hp} = \frac{\overline{T}_{ev} - \overline{T}_{cd}}{\dot{Q}_{in}} \tag{25}$$

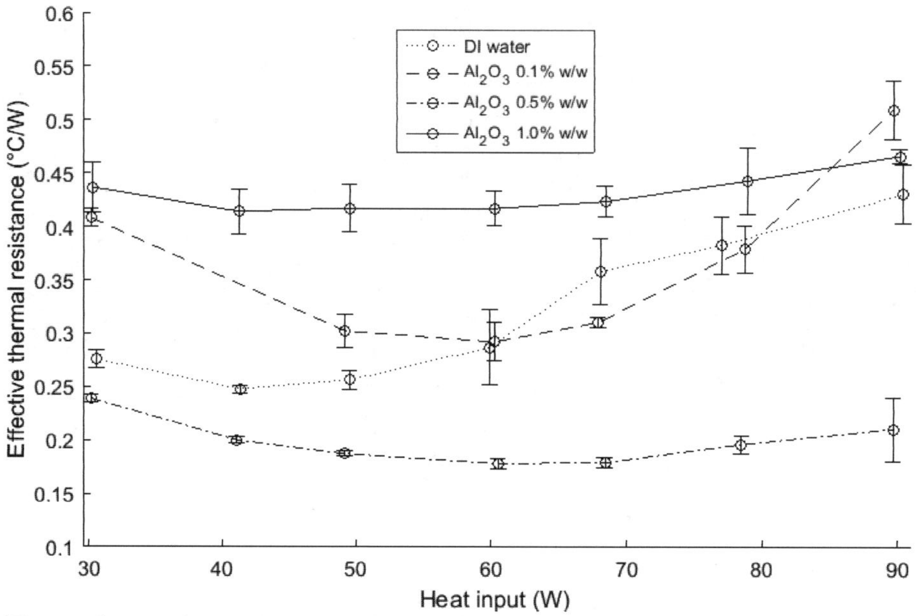

Figure 6. Experimental results of the variation of the effective thermal resistance in function of the heat input rate for different working fluids

Detection of capillary limit has been addressed to identify the heat input rate where effective thermal resistance show an increasing after a constant decrease [24]. The average thermal resistance for each working fluid is shown in Figure 6. It can be seen that experimental capillary limit with water is between 60 and 70 W, which agrees with the numerical capillary limit around 62 W. For the nanofluid 0,1% w/w, the experimental capillary limit was found between 70 and 80 W, while the numerical prediction was around 73 W. Similarly, experimental capillary limit for nanofluids 0.5% w/w and 1.0% w/w were found in the ranges 70 - 80 W and 70 – 90 W, respectively. The numerical predictions for these capillary limits were 79 W and 82 W, respectively.

As can be seen from the results, capillary limit prediction of the numerical model taking into account the deposition and agglomeration of nanoparticles was in acceptable agreement with the experimental evaluation, although exact values could not be found in the experimental conditions. Therefore, inclusion of nanoparticles proved to be effective to increase the capillary limit of the studied heat pipes. However, the relative improvement of the heat pipe performance with nanoparticles should not be attributed only to the deposition of them on the wick. The condition of the wick after the experimental tests is shown in Figure 7. These images correspond to a magnification of 1000x. It can be seen that deposition of nanoparticles was only significant for the nanofluid with concentration 1.0% w/w. With the other two nanofluids (0.1% and 0.5% w/w), few aggregates were observed in the wick, which means that the deposition in these cases is not suitable to reduce the effective pore radius significantly. However, since some nanoparticles are present in the base fluid, there are modification of physical properties as the surface tension, which affects the capillary pressure. This effect was considered in the numerical model with the simple relation in equation (17), but it is possible to think the real behavior to be more complex due to the effect of the temperature and the modification of the wick surface.

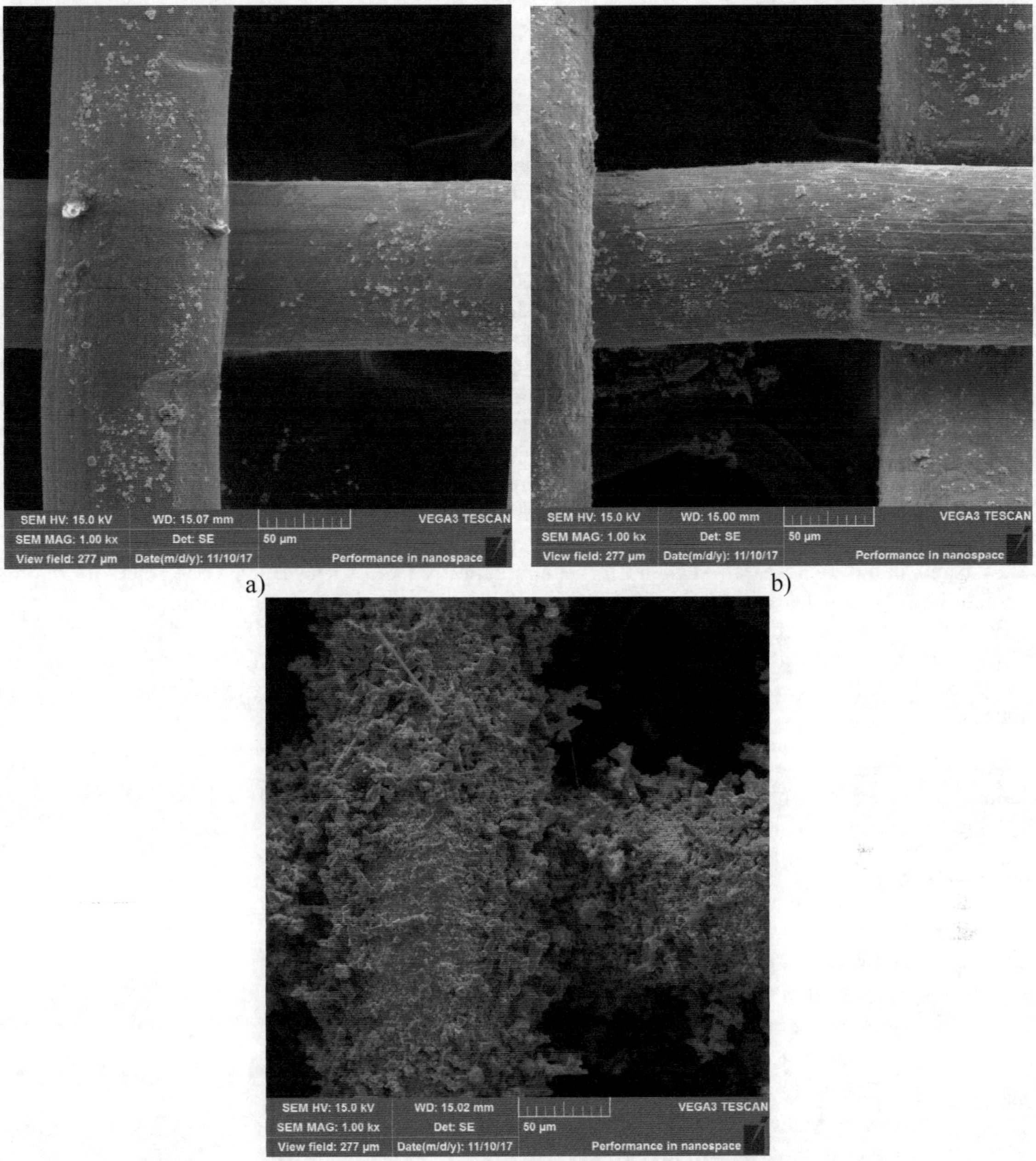

Figure 7. SEM images of the wick after experimental tests a) 0.1% w/w b) 0.5% w/w c) 1.0% w/w

Figure 6 also shows how the concentration of the nanoparticles influences the thermal performance of the heat pipes. The thermal resistance of the DI water and the nanofluid at 0.1% w/w were statistically the same in the point where the capillary limit was reached. However, the nanofluid at 0.5% showed a meaningful reduction of the effective thermal resistance in all the heat inputs. This means that addition of nanoparticles increases the heat transfer through the wick and this is likely due to the increasing of the effective thermal conductivity of the working fluid. However, further increasing in the concentration of nanoparticles leads to massive agglomeration and deposition of aggregates in the wick as shown in Figure 4 and Figure 7c. Such deposition creates and additional thermal resistance in the wick and it is expected that the effective thermal resistance of the heat pipe increase. This could be the reason which the nanofluid at 1.0% had the worst performance according to the results shown in Figure 6. Therefore, there is likely an optimum amount of nanoparticles to improve the performance of the heat pipes. In the present case, that concentration was 0.5% w/w.

4. Conclusions

A numerical model was developed to account for the effect of nanoparticle interactions when nanofluids are used as working fluid in a heat pipe. The model consisted in a coupling of a hydrodynamic model based on the Navier-Stokes equations and population balance. The numerical calculations were compared with experimental results.

Experimental results showed that stability of the nanoparticles suspensions affect the performance of the heat pipe because during the filling of the heat pipe, the early formation of aggregates in the nanofluid creates big deposition on the porous wick, blocking the circulation of the working fluid.

The numerical model predicted values of capillary limits with nanofluids up to 32% higher than that using DI water. Those predicted values were within the intervals where the capillary limit was found experimentally. According to the numerical results, the improvement in the capillary limit can be explained in the reduction of the effective pore radius which increase the capillary pressure. However, inspection of the wick after the operation of the heat pipes showed that deposition of nanoparticles was not significant at low concentrations (0.1% and 0.5% w/w), which means that the apparent decrease of the pore radius is not the main cause for improvement of the capillary pressure and other effects should be analyzed in future research such as the modification of the surface and the effect of coupled interaction of nanoparticles and temperature on properties such as the surface tension.

Finally, there was an optimum amount of nanoparticles which decreases effective thermal resistance of the heat pipes, which was 0.5% w/w in this work. Concentrations higher than that optimum concentration are suspected to increase the thermal resistance in the wick due to massive deposition of nanoparticles.

Acknowledgments

Thanks to Colciencias for financial support of the research project COL 1118-669-46092 "System for waste heat recovery with nanofluids heat pipes applied to industrial heating processes".

Nomenclature

A_c : area of the condenser (m^2)
c : volumetric concentration of nanoparticles (kg/m^3)
C_{pl} : specific heat of the liquid $(J/kg.K)$
$C_{p,nf}$: specific heat of the nanofluid $(J/kg.K)$
$C_{p,p}$: specific heat of the nanoparticle $(J/kg.K)$
C_{pv} : specific heat of the vapor $(J/kg.K)$
C_{pw} : specific heat of the wall $(J/kg.K)$
d_m : diameter of the mesh wire (m)
d_p : diameter of the nanoparticles (m)
g : acceleration of gravity (m/s^2)
h : height respect of a reference level (m)
h_c : convective coefficient $(W/m^2.K)$
h_{fg} : latent heat of vaporization (J/kg)
K : absolute permeability (m^2)
k_{eff} : effective thermal conductivity $(W/m.K)$
k_l : thermal conductivity of the liquid $(W/m.K)$
k_{layer} : thermal conductivity of the nanolayer $(W/m.K)$
k_m : thermal conductivity of the wick $(W/m.K)$
k_p : thermal conductivity of the nanoparticles $(W/m.K)$
k_w : thermal conductivity of the wall $(W/m.K)$

M	: molar mass $(kg/kmol)$
$\dot{m}_{L-V}$	: mass flow in the interface $(kg/m^3.s)$
m_i	: mass of a nanoparticle (kg)
N_i	: nanoparticles per volume of nanofluid (m^{-3})
$N_{i,r}$	: nanoparticles retained (m^{-3})
N_m	: pores per lineal meter (m^{-1})
P_0	: reference pressure (Pa)
P_C	: capillary pressure (Pa)
P_l	: pressure of the liquid in the wick (Pa)
P_i	: pressure in the interface (Pa)
P_v	: vapor pressure in the vapor core (Pa)
$\dot{Q}_{in}$	: volumetric flow of heat in the wall (W/m^3)
$\dot{Q}_{out}$	: volumetric flow of heat out of the wall (W/m^3)
R	: universal constant of gas $(8.314\ kJ/kmol.K)$
$\bar{R}$	: particular constant of gas $(kJ/kg.K)$
R_{eff}	: effective conductive thermal resistance $(W/m.K)$
R_{hp}	: effective thermal resistance of the heat pipe $(°C/W)$
r_{eff}	: effective pore radius (m)
r_p	: radius of the nanoparticle (m)
S	: mass of nanoparticles retained per mass unit of the wick (kg/kg)
S_c	: crimpage factor (1.05)
S_{max}	: max. mass of nanoparticles which can be retained per mass unit of the wick (kg/kg)
T_0	: reference temperature (K)
$\bar{T}_{cd}$	: average wall temperature across the condenser (K)
$\bar{T}_{ev}$	: average wall temperature across the evaporator (K)
T_i	: temperature in the interface (K)
T_v	: temperature in the vapor core (K)
T_w	: temperature of the wall (K)
T_∞	: temperature of the cool stream (K)
$v_{l,z}$	: axial velocity (m/s)
$v_{l,z}{}^{in}$	: inlet velocity (m/s)
$v_{l,z}{}^{out}$	: outlet velocity (m/s)
x_i	: volume of nanoparticle (m^3)
w	: thickness of the nanolayer (m)
Δz	: size of the grid (m)
ε	: porosity
ϕ	: volumetric fraction of nanoparticles
κ_a	: adsorption rate (s^{-1})
κ_d	: desorption rate (s^{-1})
μ	: viscosity of the liquid $(N.s/m^2)$
μ_l	: viscosity of the base fluid (kg/m^3)
μ_{nf}	: viscosity of the nanofluid $(N.s/m^2)$
ρ_l	: density of the liquid (kg/m^3)
ρ_{nf}	: density of the nanofluid
ρ_p	: density of the nanoparticle (kg/m^3)
ρ_s	: surface area per unit mass of the wick (m^2/kg)
ρ_w	: density of the wall (kg/m^3)
σ_l	: surface tension of the liquid (N/m)
σ_{nf}	: surface tension of the nanofluid (N/m)

References

[1] Choi S U S and Eastman J A 1995 Enhancing thermal conductivity of fluids with nanoparticles *Proc. of the 1995 ASME Inter. Mech. Eng. Congress and Exp.* (San Francisco) pp 99–105

[2] Ghanbarpour M, Nikkam N, Khodabandeh R and Toprak M S 2015 Thermal performance of inclined screen mesh heat pipes using silver nanofluids *Int. Commun. Heat Mass Transf.* **67** 14–20
[3] Taylor R a. and Phelan P E 2009 Pool boiling of nanofluids: Comprehensive review of existing data and limited new data *Int. J. Heat Mass Transf.* **52** 5339–47
[4] Kole M and Dey T K 2013 Thermal performance of screen mesh wick heat pipes using water-based copper nanofluids *Appl. Therm. Eng.* **50** 763–70
[5] Putra N, Septiadi W N, Rahman H and Irwansyah R 2012 Thermal performance of screen mesh wick heat pipes with nanofluids *Exp. Therm. Fluid Sci.* **40** 10–7
[6] Asirvatham L G, Nimmagadda R and Wongwises S 2013 Heat transfer performance of screen mesh wick heat pipes using silver–water nanofluid *Int. J. Heat Mass Transf.* **60** 201–9
[7] Shafahi M, Bianco V, Vafai K and Manca O 2010 An investigation of the thermal performance of cylindrical heat pipes using nanofluids *Int. J. Heat Mass Transf.* **53** 376–83
[8] Do K H, Ha H J and Jang S P 2010 Thermal resistance of screen mesh wick heat pipes using the water-based Al2O3 nanofluids *Int. J. Heat Mass Transf.* **53** 5888–94
[9] Shafahi M, Bianco V, Vafai K and Manca O 2010 Thermal performance of flat-shaped heat pipes using nanofluids *Int. J. Heat Mass Transf.* **53** 1438–45
[10] Do K H and Jang S P 2010 Effect of nanofluids on the thermal performance of a flat micro heat pipe with a rectangular grooved wick *Int. J. Heat Mass Transf.* **53** 2183–92
[11] Solomon A B, Ramachandran K, Asirvatham L G and Pillai B C 2014 Numerical analysis of a screen mesh wick heat pipe with Cu/water nanofluid *Int. J. Heat Mass Transf.* **75** 523–33
[12] Sadeghy R, Haghshenasfard M, Etemad S G and Keshavarzi E 2016 Investigation of alumina nanofluid stability using experimental and modified population balance methods *Adv. Powder Technol.* **27** 2186–95
[13] Zerradi H, Mizani S, Loulijat H, Dezairi A and Ouaskit S 2016 Population balance equation model to predict the effects of aggregation kinetics on the thermal conductivity of nanofluids *J. Mol. Liq.* **218** 373–83
[14] Huang Y and Chen Q 2017 A numerical model for transient simulation of porous wicked heat pipes by lattice Boltzmann method *Int. J. Heat Mass Transf.* **105** 270–8
[15] Schrage R 1953 *A Theoretical Study of Interphase Mass Transfer* (New York: Columbia University Press)
[16] Badam V K, Kumar V, Durst F and Danov K 2007 Experimental and theoretical investigations on interfacial temperature jumps during evaporation *Exp. Therm. Fluid Sci.* **32** 276–92
[17] Yu W and Choi S U S 2003 The role of interfacial layers in the enhanced thermal conductivity of nanofluids: A renovated Maxwell model *J. Nanoparticle Res.* **5** 167–71
[18] Buongiorno J 2006 Convective Transport in Nanofluids *J. Heat Transfer* **128** 240
[19] Poplaski L M, Benn S P and Faghri A 2017 Thermal performance of heat pipes using nanofluids *Int. J. Heat Mass Transf.* **107** 358–71
[20] Pak B C and Cho Y I 1998 Hydrodynamic and heat transfer study of dispersed fluids with submicron metallic oxide particles *Exp. Heat Transf.* **11** 151–70
[21] Derjaguin B and Landau L 1993 Theory of the stability of strongly charged lyophobic sols and of the adhesion of strongly charged particles in solutions of electrolytes *Prog. Surf. Sci.* **43** 30–59
[22] Verwey E J W and Overbeek J T G 1948 *Theory of the stability of lyophobic colloids* (Amsterdam: Elservier Publishing Company)
[23] Zhang T 2012 *Modeling of nanoparticle transport in porous media* (University of Texas)
[24] Wang S, Chen J, Hu Y and Zhang W 2011 Effect of evaporation section and condensation section length on thermal performance of flat plate heat pipe *Appl. Therm. Eng.* **31** 2367–73

Black box modelling of a latent heat thermal energy storage system coupled with heat pipes

A Arteconi[1,2*], F Ferracuti[2], R Tascioni[3,1], K Mahkamov[4], M Kenisarin[4], C Costa[4], L F Cabeza[5], A de Gracia[5,6], E Halimic[7], D Mullen[7], K Lynn[7] and L Cioccolanti[1]

[1]Università eCampus, via Isimbardi 10, Novedrate 22060, Italy
[2]Università Politecnica delle Marche, via Brecce Bianche 1, Ancona 60131, Italy
[3]DIAEE, Sapienza Università di Roma, via Eudossiana 18, Rome 00184, Italy
[4]Department of Mechanical Engineering and Construction, Northumbria University, Newcastle upon Tyne NE1 8ST, UK
[5]GREiA Research Group, INSPIRES Research Centre, University of Lleida, Pere de Cabrera s/n, Lleida 25001, Spain
[6]CIRIAF – Interuniversity Research Centre on Pollution and Environment "Mauro Felli", Via Duranti 63, 06125-Perugia, Italy
[7]Aavid Thermacore, Wansbeck Business Park 12, Ashington, Northumberland NE63 8QW, UK here

Abstract. This paper presents black box models to represent a LHTESS (Latent Heat Thermal Energy Storage System) coupled with heat pipes, aimed at increasing the storage performance and at decreasing the time of charging/discharging. The presented storage system is part of a micro solar CHP plant and the developed model is intended to be used in the simulation tool of the overall system, thus it has to be accurate but also fast computing. Black box data driven models are considered, trained by means of numerical data obtained from a white box detailed model of the LHTESS and heat pipes system. A year round simulation of the system during its normal operation within the micro solar CHP plant is used as dataset. Then the black box models are trained and finally validated on these data. Results show the need for a black box model that can take into account the different seasonal performance of the LHTESS. In this analysis the best fit was achieved by means of Random Forest models with an accuracy higher than 90%.

1. Introduction

Heat pipes are devices widely used for increasing heat recovery and improving thermal performance of heat exchangers in different applications [1]. In particular, heat pipes (HP) can be used to enhance the performance of LHTESS (Latent Heat Thermal Energy Storage System) which are increasingly adopted as storage systems in renewable and hybrid power plants. In literature several studies investigated heat pipes potential to increase the thermal conductivity of PCM (phase change materials) and to decrease PCM charging and discharging time [2, 3]. Robak et al. [2] found that inclusion of heat pipes increases PCM melting rates by approximately 60%. Naghavi et al. [3] shown different applications of heat pipes

* corresponding author: a.arteconi@univpm.it

coupled with PCM storage systems for low temperature and high temperature solar collectors. Among the other findings, some of the advantages guaranteed by heat pipes are: (i) very low temperature gradients between the two opposite sides of the HPs, so that an almost isothermal heat source is available for the PCM; and (ii) the heat flux through the HPs can be used to provide low heat flux densities within the LHTESS even in presence of large heat flow rates from the heat source/sink [3].

In this study a LHTESS equipped with reversible heat pipes is considered. Such system is embedded in a micro solar CHP (combined heat and power) plant for application in individual dwellings and small business residential buildings for on-site electricity and heat generation (2-kWel/18-kWth), using solar thermal energy at modest temperatures of 250-280°C (as proposed by researchers at Northumbria University [4] and under the European funded H2020 Innova MicroSolar project [5], led by Northumbria). The proposed technology is based on a linear Fresnel mirror solar concentrating collector and a micro Organic Rankine Cycle (ORC) plant [5]. In figure 1 a schematic representation of the micro solar ORC plant is provided.

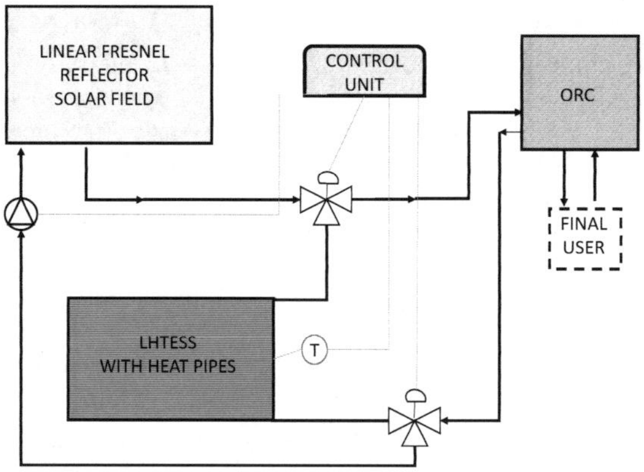

Figure 1. Schematics of the micro solar ORC pant.

The role of the storage system is to store the surplus energy from the solar field and to supply the ORC in case the solar energy is not sufficient. In order to increase the overall efficiency of the plant, in this project new reversible heat pipes are developed by Aavid Thermacore [6]. They are capable of transferring heat at high rate in both directions. This means that heat flows through the heat pipes from the solar circuit to the thermal storage tank with PCM and from the tank to the ORC plant circuit. It is expected that the proposed novel design will significantly increase the specific heat accumulation and storage efficiency. The plant will be tested on the field at Almatret (Lleida, Spain) and, in the meanwhile, a simulation model for the overall micro CHP plant was developed.

Purpose of this paper is to present the modelling of the LHTESS with the embedded heat pipes by means of an accurate but also fast computing model. At this aim, black box models are trained on the basis of a white box model simulation results, representing in detail the real system. This is intended also as a preliminary work to better understand the system behavior and the most influencing parameters, so to easily extend the methodology to the training/validation of black box models by means of experimental data collected during field test.

2. Methods

The LHTESS, as designed by Northumbria University and Aavid Thermacore [4, 6], consists of a rectangular array of heat pipes embedded in PCM and the heat transfer fluid (oil) flows through the heat pipes evaporator /condenser that is not embedded in the PCM, named oil chamber (figure.2).

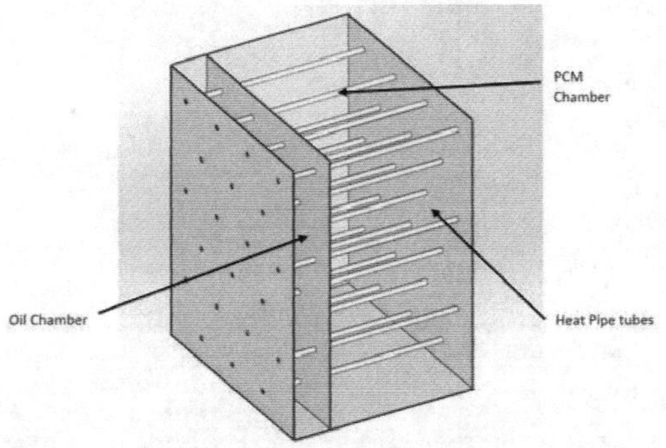

Figure 2. Schematic design of the LHTESS.

Given the complexity of the physical behavior of PCM and heat pipes system (which includes also phenomena such as hysteresis or sub-cooling), its physical representation is challenging. In order to have a good representation of the real behavior and to obtain a tool easy to implement in a comprehensive energy system simulation environment, black box models are a good option. The choice of the black box model structure and its identification procedure is a critical issue. In this work AutoRegressive models with eXogenous inputs (ARX), Nonlinear AutoRegressive networks with eXogenous inputs (NARX) and Random Forest (RF) regression models are considered. The performance of the different black box models, in the validation procedure, is compared to the results of the white box model on the basis of the Root-Mean-Square Error (RMSE) and Normalized Root-Mean-Square Error (NRMSE). In particular, accuracy of results is the key performance parameter taken into account to evaluate the effectiveness of the proposed black box models.

2.1. LHTESS and white box model description

The PCM storage material, investigated by Northumbria University and the University of Lleida [7] is made of the so-called solar salt KNO3(40wt.%)/NaNO3(60wt.%), which has high heat of fusion but low thermal conductivity; its melting temperature is in the range 216-223°C [8]. In a volume of 1.90 m3 there are 3.8 tons of this material to store about 100 kWh of latent thermal energy from the solar field in order to guarantee 4 hours of ORC unit operation during night time. Reversible heat pipes, developed by Aavid Thermacore [6], are charged with a sufficient amount of demineralized water to carry the required power; they are able to withstand a maximum pressure of 100 bar.

The PCM storage material, investigated by Northumbria University and the University of Lleida [7] is made of the so-called *solar salt* KNO_3(40wt.%)/$NaNO_3$(60wt.%), which has high heat of fusion but low thermal conductivity; its melting temperature is in the range 216-223°C [8]. In a volume of 1.90 m^3 there are 3.8 tons of this material to store about 100 kWh of latent thermal energy from the solar field in order to guarantee 4 hours of ORC unit operation during night time. Reversible heat pipes, developed by Aavid Thermacore [6], are charged with a sufficient amount of demineralized water to carry the required power; they are able to withstand a maximum pressure of 100 bar.

The physically based PCM storage tank model was developed according to the guidelines of the IEA Task 32 report on advanced storage concepts [9]. The PCM is supposed isotropic and isothermal in each time-step. Hysteresis and sub-cooling effects are neglected. The white box model represents the behavior of the heat pipes by means of: (i) a limitation in the maximum power exchanged with the oil because of the limited heat pipes capacity (40 kW) and (ii) a minimum temperature difference between the oil and the PCM equal to 5°C.

The dataset used for training the black box models is represented by the simulation results of the white box model. The white box model is implemented in TRNSYS (an environment for transient representation of thermal systems behavior [10]) within the overall micro-solar CHP plant. A full year of operation of the plant, set in the city of Cagliari in Italy (local coordinates 39° 22' 38.41'' N and 9° 12' 16.61'' E), is considered. Therefore, the operating conditions of the LHTESS during its operation in the integrated plant are taken into account. The simulation time step is 10 minutes (T_c, sampling period).

The input variables to the black box models are:

- Oil inlet temperature (T_{OIL}^{in});
- Oil inlet flow rate (F_{OIL}^{in});
- Ambient temperature (T_A);
- Direct Normal Irradiance (DNI) of the sun;
- Operation mode of the plant (OM). Indeed, there are 6 different operation modes in the control strategy of the micro solar CHP plant, where the LHTESS can be charged or discharged in different operating conditions.

The input variables are averaged to better reflect their influence on the output variables by a test and trial procedure. Eventually, T_A is filtered by a median filter over a 24 hours period, while T_{OIL}^{in}, F_{OIL}^{in} and DNI are filtered by a median filter over a 2 hours period and OM is filtered by a mode filter over a 24 hours period. The output variables of the black box models, i.e. the variables to be predicted, are:

- Oil outlet temperature (T_{OIL}^{out});
- Temperature of the TES (T_{TES});
- Thermal losses of the TES to the environment (P_{LOSS}).

In figure 3 the input and output variables of the black box models are represented.

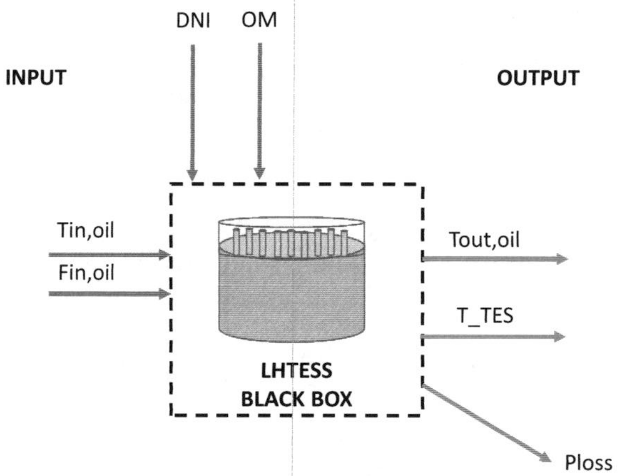

Figure 3. Schematic representation of the Black Box models input and output variables.

2.2. *ARX models*

The autoregressive model with exogenous inputs is a linear model defined by the equation 1 as in [11]:

$$A(z)y(k) = B(z)u(k) + v(k) \tag{1}$$

with $A(z)$ and $B(z)$ denoting polynomials with respect to time-shift operator z:

$$A(z) = 1 + a_1 z^{-1} + \cdots + a_{na} z^{-na} \tag{2}$$

$$B(z) = b_0 + b_1 z^{-1} + \cdots + b_{nb} z^{-nb} \tag{3}$$

The ARX model considered in this study for simulation can be mathematically represented as:

$$\sum_{i=0}^{n_a} a_i\,\hat{y}(k-i) = \sum_{j=0}^{n_b} b_{1j}\,F_{OIL}^{in}(k-j-n_k) +$$
$$+\textstyle\sum_{j=0}^{n_b} b_{2j}\,T_{OIL}^{in}(k-j-n_k) + \sum_{j=0}^{n_b} b_{3j}\,T_A(k-j-n_k) + \sum_{j=0}^{n_b} b_{4j}\,DNI(k-j-n_k) +$$
$$\textstyle\sum_{j=0}^{n_b} b_{5j}\,OM(k-j-n_k) \quad (4)$$

where $k = nT_c$ denotes time, $n \in Z$, sampling period T_c and a_i, b_{1j}, b_{2j}, b_{3j}, b_{4j} and b_{5j} denote the unknown model parameters. n_a, n_b and n_k are the orders of the model (so called hyper-parameters) associated with output, inputs and dead time in the system, respectively. $\hat{y}(k)$ represents the estimated outputs, namely T_{OIL}^{out}, T_{TES} and P_{LOSS}.

2.3. NARX models

The A nonlinear autoregressive network with exogenous inputs is a special case of Recurrent Neural Network (RNN), which proved to be a very successful modeling tool for nonlinear systems and especially time series [12]. NARX networks converge faster, are less likely to develop long-term dependencies and typically have better generalization abilities than other networks [12]. The NARX structure considered in this study for simulation can be mathematically represented as:

$$\hat{y}(k) = f\big[F_{OIL}^{in}(k-n_k), \dots, F_{OIL}^{in}(k-n_k-n_b),\ F_{OIL}^{in}(k-n_k), \dots, F_{OIL}^{in}(k-n_k-n_b), T_A(k-n_k), \dots, T_A(k-n_k-n_b), DNI(k-n_k), \dots, DNI(k-n_k-n_b), OM(k-n_k), \dots, OM(k-n_k-n_b)\big] \quad (5)$$

The nonlinear mapping f is generally unknown and can be approximated, for example, by a standard Multilayer Perceptron (MLP) network.

Since in this paper the black box model is intended to be used as simulation model, the true past values of $\hat{y}(k)$ are not available when the model is deployed (differently from predictive models).

Therefore, the resulting architecture is a particular case of NARX network.

2.4. Random forest

Breiman [13] defined a "random forest" as a variant of bagging meta-algorithm. Random forest is a general class of ensemble building methods which use decision trees for regression or classification. To be labeled a "random forest", an ensemble of decision trees should be built by generating independent and identically distributed random vectors and should make use of each vector to model a decision tree. Therefore, a random forest could be built by sampling from the data set or feature set or just varying randomly some of the parameters of the tree. Any combination of these sources of diversity will also lead to a random forest.

2.5. Identification procedure

Data pre-processing is the first step before the identification procedure. In particular, for ARX, data detrend is performed. This data processing operation helps to estimate more accurately linear models and overcome their inability to capture arbitrary differences between input and output signal levels. For NARX, data are normalized in the range [-1; 1] since a sigmoid function is considered in the hidden layer. In order to reduce the overfitting issue, regularization term is considered for both ARX and NARX. For all the three models, the 10-fold cross-validation technique is used. Only for RF modelling, the input operation mode of the plant (OM) is considered as a categorical predictor.

The System Identification Toolbox in Matlab [14] is applied on a training dataset to estimate the unknown parameters for ARX. In the NARX model, instead, the optimal values of the model parameters have been estimated using the Neural Network Toolbox. Finally, RF regression model is built by the Statistics and Machine Learning Toolbox.

A grid search optimization procedure is considered to set the hyperparameters of the three models, in particular, the orders of ARX (i.e, n_a, n_b and n_k), the orders of NARX (i.e, n_b and n_k) and the orders of regression (i.e., n_b and n_k), the leaf size and the number of decision trees of RF model.

After identification, the models are tested by means of the following indicators:

- the RMSE-values on the validation dataset;
- the level of fit (NRMSE) on the whole dataset.

The RMSE (root mean square error) is the objective function to be minimized both for training and validation sets. RMSE is defined as follows:

$$RMSE = \left(\frac{1}{N}\sum_{k=1}^{N} e(k)^2\right)^{1/2} \tag{6}$$

where N is the number of samples and $e(k) = y(k) - \hat{y}(k)$ represents the difference between the variable $\hat{y}(k)$ predicted by the simulation model and the variable $y(k)$ predicted by the white box model. NRMSE values are the normalized root mean square errors and thus they show in percentage the model goodness of fit.

3. Results

The different black box models were first trained and then tested, as described in the previous section. Firstly, the ARX models were considered, because they have a simpler structure. Then the NARX and eventually the RF models, in order to improve the performance in fitting with the white box model.

3.1. ARX model results

The optimum parameters of the ARX model identified are: n_a=4, n_b=4 and n_k=3 for T_{TES}; n_a=1 , n_b=9 and n_k=0 for T_{OIL}^{out}; n_a=0 , n_b=6 and n_k=0 for P_{LOSS}. The RMSE and NRMSE values are reported in Table 1.

Table 1. ARX results.

Output variable	RMSE	NRMSE
T_{TES}	3.1 (°C)	0.241
T_{OIL}^{out}	8.4 (°C)	0.914
P_{LOSS}	71.4 (kJ/h)	0.362

The NRMSE obtained with the ARX model is not very good, especially for the variable T_{TES}. Looking in detail to the trend of such variable during the year (figure 4a e figure 4b), a different behavior of the storage system in summer and in winter is evident. That was expected, because the micro-solar CHP plant performance is strongly influenced by the solar radiation available. In winter the plant is off for longer periods of time and the LHTESS is partially charged, without reaching the melting condition, and rarely discharged. In summer, instead, the LHTESS works properly and can be fully charged and discharged using the latent heat available.

The ARX model, as here identified, cannot represent such different behaviors by means of a single model. Looking at the predicted and original values of the output variables (figure 4a e figure 4b), the model can mimic the LHTESS temperature much better in summer than in winter. For this reason, a separate identification procedure for the two seasons was performed. The global performance indicators of the two new ARX models are reported in Table 2.

Table 2. ARX results in summer and winter.

Output variable	NRMSE summer	NRMSE winter
T_{TES}	0.455	0.077
T_{OIL}^{out}	0.914	0.884
P_{LOSS}	0.415	0.167

The results show an increased ability, even if still low, of the summer ARX model to predict the output variables trend, while in winter the goodness of fit is still poor.

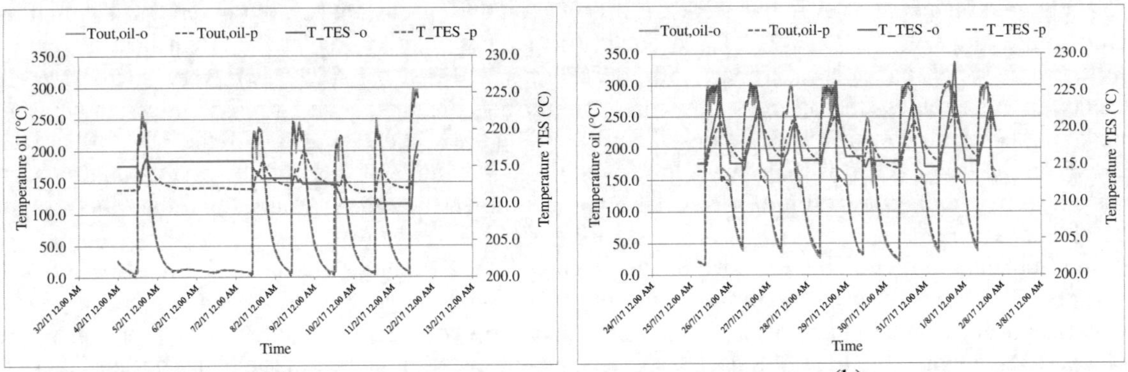

(a) (b)

Figure 4. Predicted (p) vs original (o) values of the output variables for the ARX model in (a) winter and in (b) summer.

3.2. NARX model results

Given the unsatisfying results shown in the previous section, another class of black box models was considered. The optimum parameters of the NARX model identified are: n_b=12 and n_k=0.
The RMSE and NRMSE values are reported in Table 3. Again, a single NARX model cannot properly predict well the different LHTESS behavior in the different seasons. Considering the results already obtained for the ARX model, the NARX model has not been furtherly investigated.

Table 3. NARX results.

Output variable	RMSE	NRMSE
T_{TES}	2.8 (°C)	0.303
T_{OIL}^{out}	6.9 (°C)	0.931
P_{LOSS}	60.4 (kJ/h)	0.463

3.3. RF model results

In order to improve the ability to reproduce the behavior of the LHTESS during the whole year, a more complex category of black box models was considered. Indeed the RF models contain internal different models and their outcome is an average of all the included models. The optimum parameters of the RF model identified are: n_b=12 and n_k=0 and it is used 1 leaf size and 20 decision trees. In Table 4 the corresponding RMSE and NRMSE values are reported.

Table 4. RF results.

Output variable	RMSE	NRMSE
T_{TES}	0.7 (°C)	0.904
T_{OIL}^{out}	4.2 (°C)	0.973
P_{LOSS}	15.3 (kJ/h)	0.929

In this case the NRMSE (FIT) obtained is pretty good, especially for T_{OIL}^{out} and P_{LOSS}. Even the prediction of T_{TES} has been greatly improved and can be considered acceptable. However, looking at the RMSE values, it must be highlighted that, even if the absolute value of the RMSE is low, it could have a huge influence. For example, a RMSE of 0.7°C on the T_{TES} is not big, but given that the melting temperature range is 7°C (216-223°C), it could affect the ability to predict the latent heat available. Furthermore, T_{OIL}^{out} has values around 250-300°C, thus a RMSE of 4.2°C is a very small error in prediction, nevertheless the oil mass flow rate is pretty big, then also a small error in this temperature prediction is translated in a consistent mistake in assessing the heat exchanged with the TES. Instead the RMSE on the P_{LOSS} (kJ/h) is really very low, even if the relevance of this output variable is limited. It could also be obtained as result of a calculation, once known the input variables and the other output variables.

In figure 5a and figure 5b the predicted and original output variables are shown for the two seasons. Predicted T_{OIL}^{out} follows pretty good the trend of the original value, while T_{TES} shows, especially in winter, an oscillating trend not visible in the original variable. As already mentioned, in winter there are long periods in which the LHTESS does not exchange energy with the rest of the plant, because there is not a surplus of energy to be stored or recovered. During these periods the TES exchanges heat only with the external environment as thermal losses. Thus, the T_{TES} results highly correlated with the external ambient temperature and this is reflected by the prediction of the black box model. However, the variability of the external ambient temperature is much higher than the actual variability of the temperature of the storage tank.

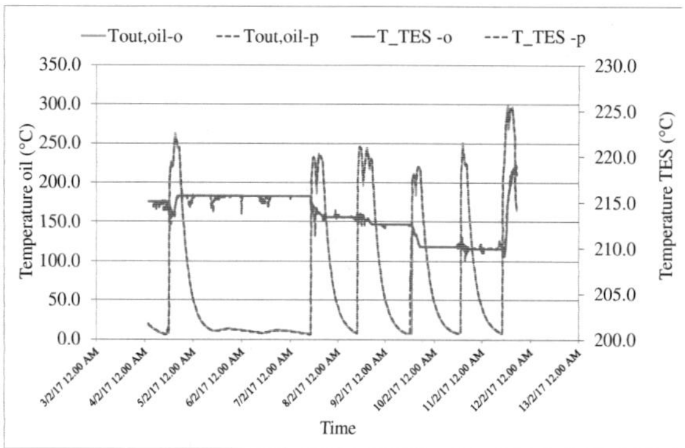

(a)

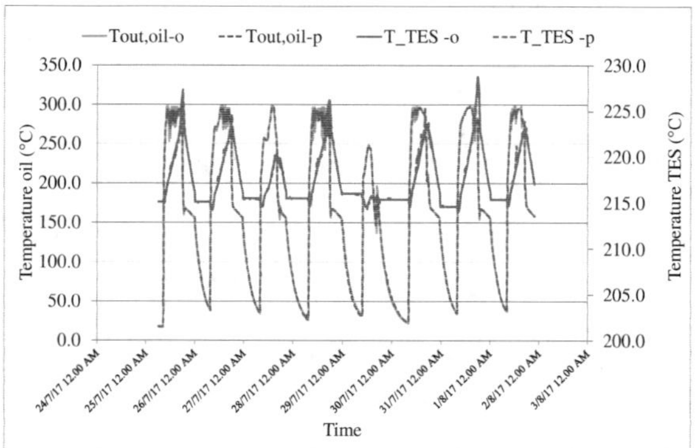

(b)

Figure 5. Predicted (p) vs original (o) values of the output variables for the RF model in (a) winter and in (b) summer.

Such aspect could be further improved with an in depth study of the influence of the input data on the output variables, in order to better represent the system behavior during the whole year. In Fig. 6 the influence of each input variable on the output variables is shown. In particular, for any variable, the prediction importance is the increase in prediction error if the values of that variable are permuted across the out-of-bag observations. This measure is computed for every tree, then averaged over the entire ensemble and divided by the standard deviation over the entire ensemble (it is a number >0).

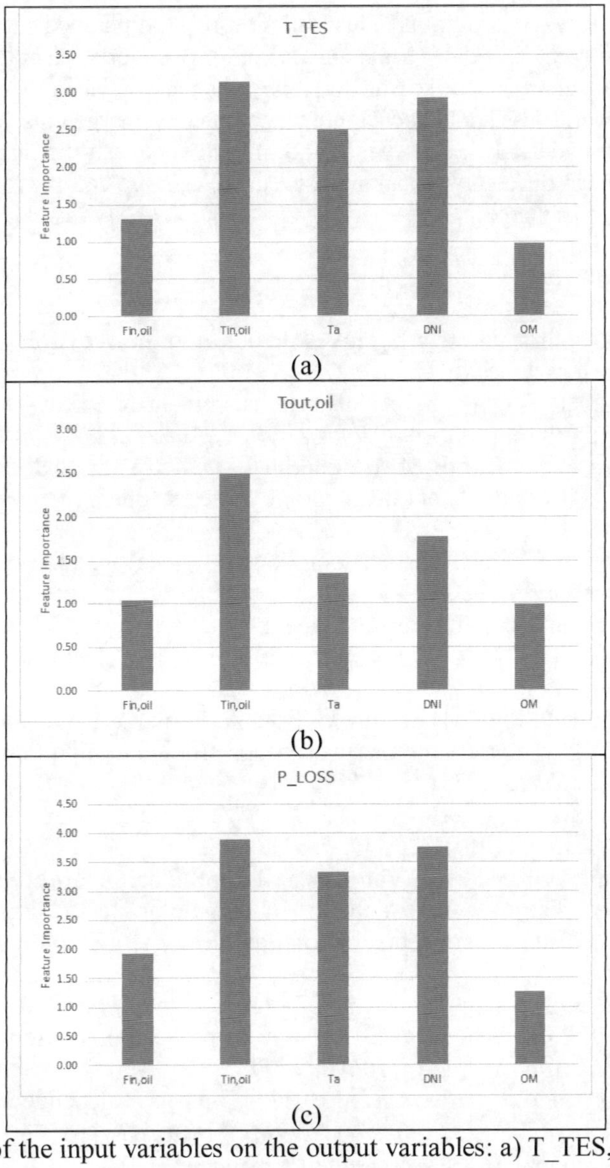

Figure 6. Influence of the input variables on the output variables: a) T_TES, b) Tout,oil, c) Ploss.

It can be observed that T_{OIL}^{in} and DNI are the most influencing variables for all the different outputs. Moreover it is confirmed that the ambient temperature has also a quite important effect on the heat losses and on the TES temperature, as already discussed previously.

4. Conclusions

In this study a first attempt to model a complex Latent Thermal Heat Energy Storage System (LHTESS) coupled with heat pipes by means of black box models is presented. Several configurations of black box models were considered in order to find a good fit between predicted and original values of the output variables of the system. As original values were taken the simulation results of a white box model of the LHTESS placed in a micro-solar CHP plant, obtained with a dynamic simulation tool to represent the system behavior during variable operating conditions.

The analysis performed is intended as a preliminary study to better understand the influence of the input variables of the system on its outputs. This will simplify the planned future work of training the black box models on real measured data. Main conclusion of this work is that the HP-LHTESS shows a very different behavior with changing the season and it is particularly difficult to predict its behavior during winter, when the system works less hours and cannot be completely charged/discharged for lack of surplus of energy. Thus, a single model for the all year cannot mimic very well the output variables. Therefore, different models identified in different seasons must be used, or more complex models, as the RF model here proposed, need to be considered.

5. References

[1] Amini A, Miller J, Jouhara H 2017 An investigation into the use of the heat pipe technology in thermal energy storage heat exchangers, *Energy* **136** 163-172

[2] Robak C W, Bergman T L, Faghri A 2011 Enhancement of latent heat energy storage using embedded heat pipes, *International Journal of Heat and Mass Transfer* **54** 3476–3484

[3] Naghavi M S, Ong K S, Mehrali M, Badruddin I A, Metselaar H S C 2015 A state-of-the-art review on hybrid heat pipe latent heat storage systems. Energy Conversion and Management **105** 1178-1204

[4] Northumbria University | Newcastle upon Tyne | Study in the Best Student City n.d. *https://www.northumbria.ac.uk/*

[5] Innovative Micro Solar Heat and Power System for Domestic and Small Business Residential Buildings, project ID:723596, *http://innova-microsolar.eu/*, last visited on 13/10/2017.

[6] Thermacore n.d. *http://www.thermacore-europe.com/*

[7] Maldonado J M, Fullana-Puig M, Martin M, Solé A, Fernández A G, de Gracia A, Cabeza L F 2018 Phase change materials for thermal energy storage at high temperature range between 201°C and 270°C. *Energies* **11**(4), 861

[8] Serrano-López R, Fradera J, Cuesta-López S 2013 Molten salts database for energy applications. *Chem Eng Process*. doi:10.1016/j.cep.2013.07.008.

[9] Gantenbein P, Jaenig D, Kerskes H, Van Essen M 2008 Final report of Subtask B; Chemical and Sorption Storage; The overview A Report of IEA Solar Heating and Cooling programme - Task 32 Advanced storage concepts for solar and low energy buildings Report B7 of Subtask B

[10] Welcome | TRNSYS : Transient System Simulation Tool n.d. *http://www.trnsys.com/*

[11] Smrekar J, Potocnik P, Senegacnik A 2013 Multi-step-ahead prediction of NOx emissions for a coal-based boiler, *Applied Energy* **106** 89 – 99

[12] Lin T., Horne B.G., Tino P., Giles C.L., Learning long-term dependencies in NARX recurrent neural networks, *IEEE Transactions on Neural Networks* 7 (1996) 1329–1338.

[13] Breiman, L. 2001 Random Forests *Machine Learning* **45** 5–32

[14] MATLAB - MathWorks n.d. *https://www.mathworks.com/products/matlab.html*

Aknownledgement

This study is a part of the Innova MicroSolar Project, funded in the framework of the European Union's Horizon 2020 Research and Innovation Programme (grant agreement No 723596). Prof. Cabeza would like to thank the Catalan Government for the quality accreditation given to their research group (2017 SGR 1537). GREA is certified agent TECNIO in the category of technology developers from the Government of Catalonia. Dr. Alvaro de Gracia has received funding from the European Union's Horizon 2020 research and innovation programme under the Marie Sklodowska-Curie grant agreement No 712949.

IOP Conf. Series: Materials Science and Engineering 1139 (2021) 012013 doi:10.1088/1757-899X/1139/1/012013

Thin film evaporation model for two-phase capillary heat transfer devices: examination of boundary conditions and vapour pressure gradient

S Ahmed[1] and M Pandey[1*]

[1]Department of Mechanical Engineering, Indian Institute of Technology Guwahati, Assam, 781039, India

Abstract. Thin film evaporation model has been used by several researchers to study the transport phenomena of two-phase capillary devices, such as heat pipes and capillary pumped loops. The present work is focused on mathematical modeling of the evaporation phenomena in such devices. Liquid-vapour interface in the evaporator is modelled using a thin film model and the lubrication approximation, along with a slip boundary condition at the wall and a shear boundary condition at the interface. Different models are used for the evaporating mass flux and the vapour pressure gradient at the liquid-vapour interface is considered. An attempt is also made to determine the non-evaporating film thickness that satisfies the underlying physics. The film thickness profile and the pressure components are obtained by numerical simulations. It is found that the choice of the evaporating mass flux model has a significant effect on the results, and is very important for heat transfer characterization.

1. Introduction

Two-phase Capillary Devices are those devices which use the capillary force to pump working fluid along their system. These devices work independently of gravity and their performance is not affected by their orientation. Examples of such devices include those which come under heat pipe family and capillary pumped loop. Their components consist of an evaporator, condenser, wick (to provide capillary support) and transport channels. Heat is applied in the evaporator where the working fluid turns into vapor and reaches the condenser via transport channels. In condenser, heat is rejected and the vapor turns into liquid which is then returned to the evaporator via the wick. The wick consists of either porous material (e.g. sintered metal powder) or microchannels. When the heated working fluid wets the wall of these channels, the extended meniscus (figure 1) can be divided into three regions:

* corresponding author: manmohan@iitg.ac.in

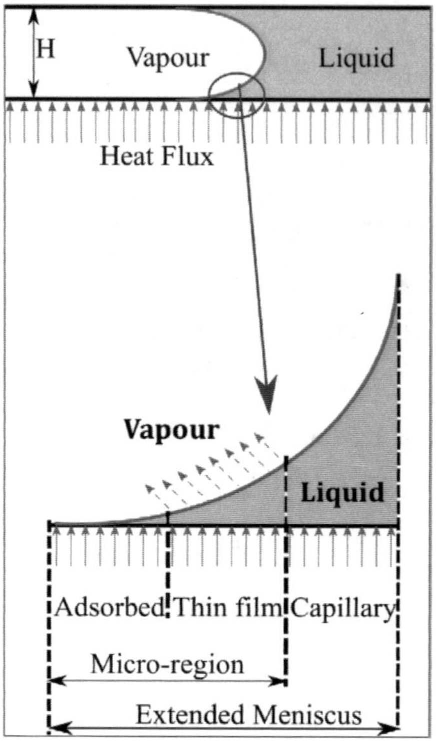

Figure 1. Extended meniscus

- **The capillary region** where the capillary force dominates. It is also called intrinsic meniscus.
- **The thin film region** where maximum amount of heat transfer occurs due to low thermal resistance. It is dominated by both capillary and disjoining pressure.
- **The adsorbed region** (also known as equilibrium thin film region) where no evaporation takes place due to high disjoining pressure.

Dergajuin laid down the foundation stone for the study of ultra-thin films. He defined the concept of disjoining pressure as "The difference between the pressure in a region of a phase adjacent to a surface confining it, and the pressure in the bulk of this phase".Since then, researchers have carried out both experimental and theoretical studies. Among them, the work of Wayner Jr. and Coccio [1] deserves special mention as it was in their experimental study, that they postulated the existence of a thin film adjacent to the intrinsic meniscus which has been later verified by several researchers. Wayner et al. [2] further simplified the numerical model by implementing the lubrication approximation and simplifying the Schrage correlation for calculating interfacial mass transfer. This later became popularly known as Wayner's Model. Wang et al. [3] studied heat transfer characteristics of both thin film and micro region. They used both Schrage model and Wayner model in the calculation of evaporating mass flux. They concluded that thin film profile generated by Wayner model agrees with that of Schrage model upto superheat of 5 K. However, for high superheat, Schrage model has to be used. Apart from that, they also studied the effect of capillary pressure in their role in suppression of evaporation. Hanchak et al. [4] implemented a better disjoining pressure model which was based on a retarded van der Waals interaction. They also carried out experimental verification of their numerical model. They applied reflectometry to measure the thickness of thin film of n-octane on silicon wafer. Jasvanth et al. [5] carried out a comparative study of evaporating thin film for different working fluid in a loop heat pipe. They confirmed that heat transfer coefficient of a loop heat pipe directly depends on fluid figure of merit (namely interline heat flow parameter [6–8] and heat pipe figure of merit, shown by equation (1)).

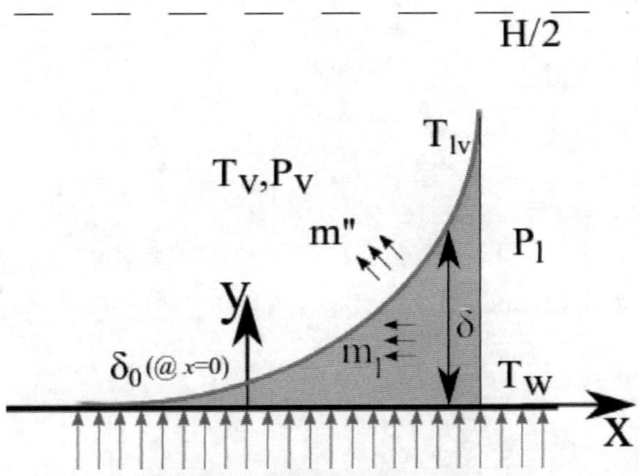

Figure 2. Thin film region

$$\mathrm{FoM_{IL}} = \frac{\rho_l \bar{A}\, \lambda}{\mu_l} \qquad (1a)$$

$$\mathrm{FoM_{HP}} = \frac{\sigma \rho_l\, \lambda}{\mu_l} \qquad (1b)$$

Du and Zhao [9] worked on getting more accurate solution by taking improved boundary conditions. Later on, Akkus and Dursunkaya [10] extended their work by starting the calculation from intrinsic meniscus. In another work, Akkus et al. [11] developed a two dimensional solution methodology for the thin film model. Fu et al. [12] considered the bulk flow between the vapour and interface phase. They directly solved the momentum conservation equation instead of lubrication approximation in their analysis. They also studied the effect of accommodation coefficient and concluded that it significantly affects the important parameters like interfacial temperature, heat flux, mass flow rate etc. but has no effect on the thin film profile.

Although researchers came up with numerous different strategies to improve the model, they had one thing in common. They assumed uniform vapour pressure, and, as a consequence, ignored the vapour pressure gradient. However, Park et al. [13] had considered the vapour pressure gradient and proved that it affects the thickness of thin film. They also incorporated slip boundary condition in their model.

In the present study, effect of vapour pressure gradient is investigated using different evaporation flux model. Apart from that, issues associated with boundary conditions is examined and a simple yet effective solution is put forward. Thus, this work can be considered an extension of Wang et al. [3], Park et al. [13], and Kou et al. [14] and lays down foundation stone for our future work [15, 16].

2. Mathematical model

Figure 2 shows the thin film region in details. To simplify the model, following assumptions are made:

i. Two-dimensional and steady-state laminar flow
ii. Pressure is a function of x-coordinate only.
iii. Incompressible liquid and vapour state
iv. Uniform wall temperature
v. Working fluid having constant properties and surface tension
vi. Negligible convective terms in the momentum equation
vii. Evaporation is restricted to thin film region only

2.1. Governing equations

In the thin film, both capillary and disjoining force act. Thus, the pressure difference is balanced out by both capillary and disjoining forces, which is shown in equation (2).

$$P_v - P_l = P_c + P_d \qquad (2)$$

Here

$$P_c = \sigma \frac{\frac{d^2\delta}{dx^2}}{\left\{1 + \left(\frac{d\delta}{dx}\right)^2\right\}^{3/2}} \quad (3)$$

$$P_d = \frac{\bar{A}}{\delta^3} \quad (4)$$

Differentiating equations (2), (3) and (4) the following equation is obtained:

$$\frac{d^3\delta}{dx^3} = \frac{3\left(\frac{d\delta}{dx}\right)\left(\frac{d^2\delta}{dx^2}\right)^2}{1 + \left(\frac{d\delta}{dx}\right)^2} + \frac{1}{\sigma}\left(\frac{dP_v}{dx} - \frac{dP_l}{dx} + 3\frac{\bar{A}}{\delta^4}\frac{d\delta}{dx}\right) \quad (5)$$

The equation for liquid pressure gradient (6) and vapour pressure gradient (7) is obtained from the lubrication approximation and momentum equation respectively:

$$\frac{dP_l}{dx} = \mu_l\left(\frac{d^2u_l}{dy^2}\right) \quad (6)$$

$$\frac{dP_v}{dx} = -\mu_l\left(\frac{d^2u_v}{dy^2}\right) \quad (7)$$

Equations (6) and (7) are integrated. Boundary conditions used are:

- **For equation (6):** Here, shear and slip boundary conditions are applied.
    i. $At\ y = \delta,\ \frac{du_l}{dy} = \frac{\tau_i}{\mu_l}$
    ii. $At\ y = 0,\ u_l = \beta\frac{du_l}{dy}$

  Here $\beta = 5 \text{ x } 10^{-9}m$ [13].
- **For equation (7):**
    i. $At\ y = H/2,\ \frac{du_v}{dy} = 0$
    ii. $At\ y = \delta,\ u_v = u_\delta$

This gives:

$$u_l = \frac{1}{2\mu_l}\frac{dP_l}{dx} + \frac{\tau_i}{\mu_l}(y + \beta) \quad (8)$$

$$u_v = u_\delta - \frac{1}{2\mu_v}\frac{dP_v}{dx} + (y + \delta - H) \quad (9)$$

The value of uδ can be found out from equation (8). Now, we know that:

$$\dot{m}_l = \int_0^\delta \rho_l u_l dy \quad (10)$$

$$\dot{m}_v = \int_0^{H/2} \rho_v u_v dy \quad (11)$$

Substituting the values of u_l and u_v in equations (10) and (11) respectively:

$$\frac{dP_l}{dx} = -\frac{3\nu_l}{\delta^2(\delta + 3\beta)}\left\{\dot{m}_l - \frac{\tau_i\delta}{\nu_l}\left(\frac{\delta}{2} + \beta\right)\right\} \quad (12)$$

$$\frac{dP_v}{dx} = -\frac{24\nu_v}{H^3 + 6H\delta(\delta - H)}\left\{\dot{m}_v - \rho_v u_\delta\frac{H}{2}\right\} \quad (13)$$

Now at the liquid-vapour interface, the shear stress for vapour is equal to the shear stress for liquid.

$$\therefore\ \tau_i = \mu_l\frac{\partial u_l}{\partial y}\bigg]_{y=\delta} = \mu_v\frac{\partial u_v}{\partial y}\bigg]_{y=\delta}$$

Table 1. List showing different values of first and second derivative of δ from literature

Paper	$\frac{d\delta}{dx}$	$\frac{d^2\delta}{dx^2}$
[13]	0	0
[18]	0	5 x 10^{-6}
[4]	0	Far-field boundary condition
[3],[14]	slightly greater than zero	Far-field boundary condition

Substituting τ_i at equations (12) and (13),

$$\frac{dP_l}{dx} = -\frac{3v_l}{\delta^2(\delta+3\beta)}\left\{\dot{m}_l + \frac{\delta}{v_l}\left(\frac{\delta}{2}+\beta\right)\left(\delta - \frac{H}{2}\right)\frac{dP_l}{dx}\right\} \quad (14)$$

$$\frac{dP_v}{dx} = -\frac{24v_v}{H^3 + 6H\delta(\delta - H)}\left[\dot{m}_v + \rho_v\frac{H}{2}\left\{\frac{1}{2\mu_l}\delta(\delta+2\beta)\right\}\frac{dP_l}{dx} + \frac{1}{2\mu_l}(\delta+\beta)(2\delta - H)\frac{dP_v}{dx}\right] \quad (15)$$

Mass flow rates of liquid and vapour are given by:

$$\dot{m}'' = -\frac{d\dot{m}_l}{dx} = \frac{d\dot{m}_v}{dx} \quad (16)$$

Two different correlations are used to evaluate $\dot{m}''$:

- **Linear Model:**

$$\dot{m}'' = \frac{q''}{\lambda} \quad (17)$$

- **Wayner Model** (from Wayner et al.[2]):

$$\dot{m}'' = \frac{2\hat{\sigma}}{2-\hat{\sigma}}\left(\frac{\bar{M}}{2\pi\bar{R}}\right)^{1/2}[a(T_{lv} - T_v) - b(P_c + P_d)] \quad (18)$$

Here, the working fluid considered is a non-polar liquid. Thus, σ^ is taken as unity. Correlation of the other parameters are given below:

$$T_{lv} = T_W - \frac{\delta\dot{m}''\lambda}{k_l} \quad (19)$$

$$a = \left(\frac{P_v\bar{M}\lambda}{\bar{R}T_vT_{lv}^{\frac{3}{2}}}\right) \quad (20)$$

$$a = \left(\frac{V_lP_v}{\bar{R}T_{lv}^{3/2}}\right) \quad (21)$$

Equation (19) is the same equation which Moosman and Homsy [17] used to modify the Wayner's Model by eliminating T_{lv}.

2.2. *Boundary conditions*

So, the equations to be solved are: (5), (14), (15) and (16). Boundary conditions (@x = 0) consists of:

$$\delta = \delta_0, \quad \frac{d\delta}{dx} = 0, \quad \frac{d^2\delta}{dx^2} = 0, \quad \dot{m}_l = 0, \quad \dot{m}_v = \dot{m}_{l(x=L)}, \quad P_v = P_{sat},$$

 doi:10.1088/1757-899X/1139/1/012013

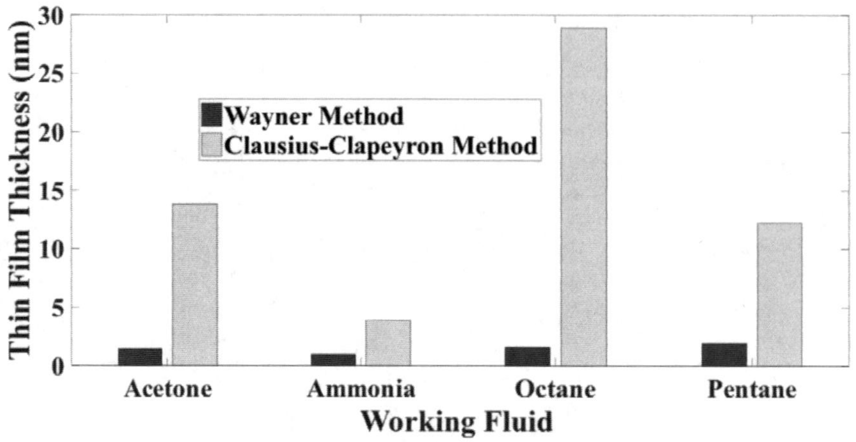

(a)

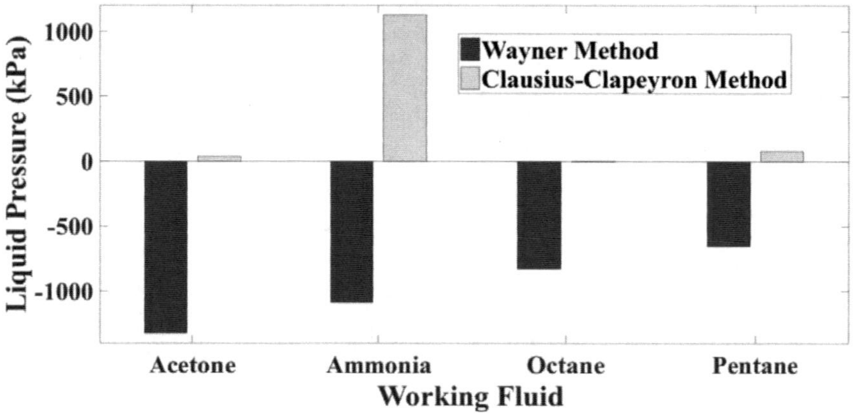

(b)

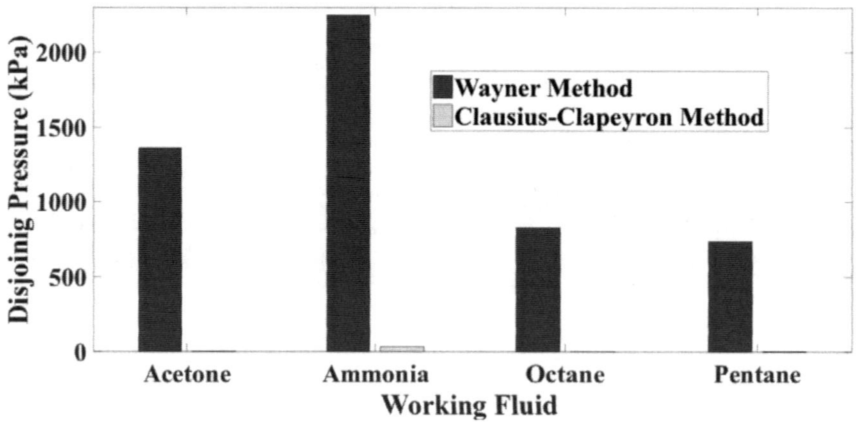

(c)

Figure 3. Comparision of: (a) thin-film thickness, (b), Liquid Pressure and (c) Disjoining Pressure, at x = 0 calculated from 'Wayner' and 'Clausius-Clapeyron' method for different working fluid at same saturation temperature and superheat

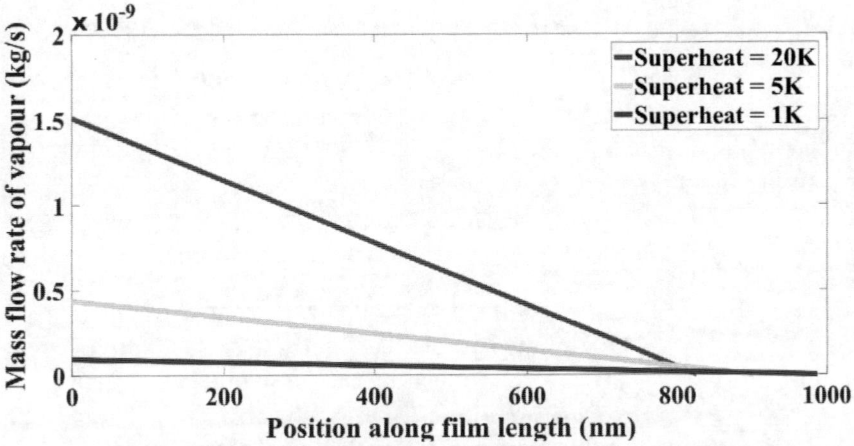

Figure 4. Effect of superheat on the mass flow rate of vapour

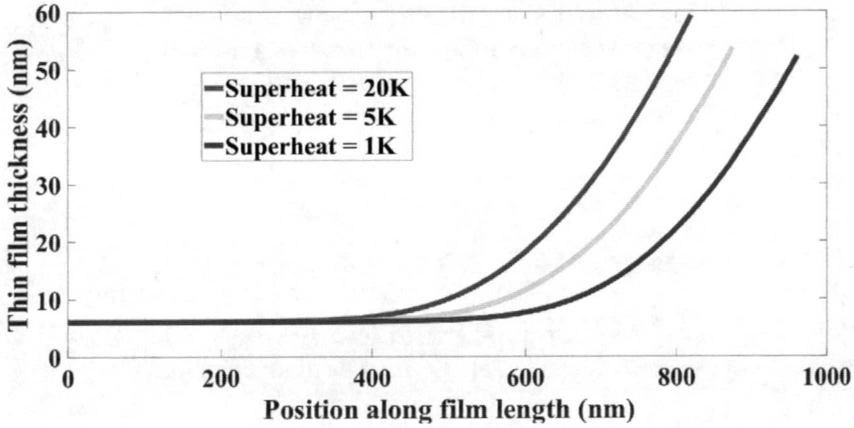

(a)

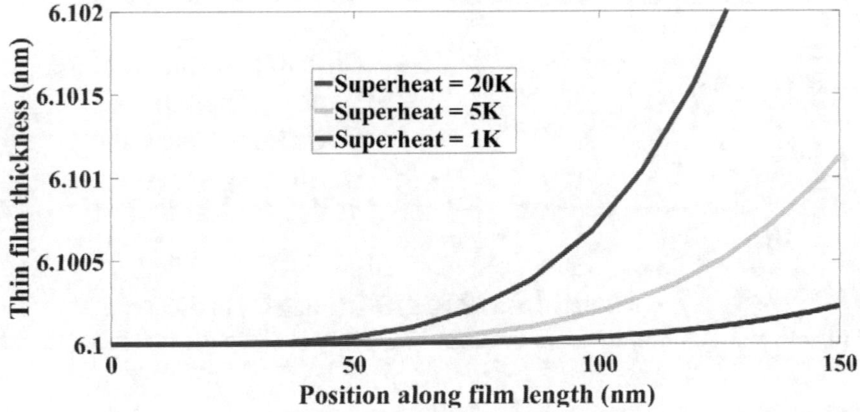

(b)

Figure 5. Effect of superheat on the thin-film thickness: (a) Overall variation, (b) Initial variation

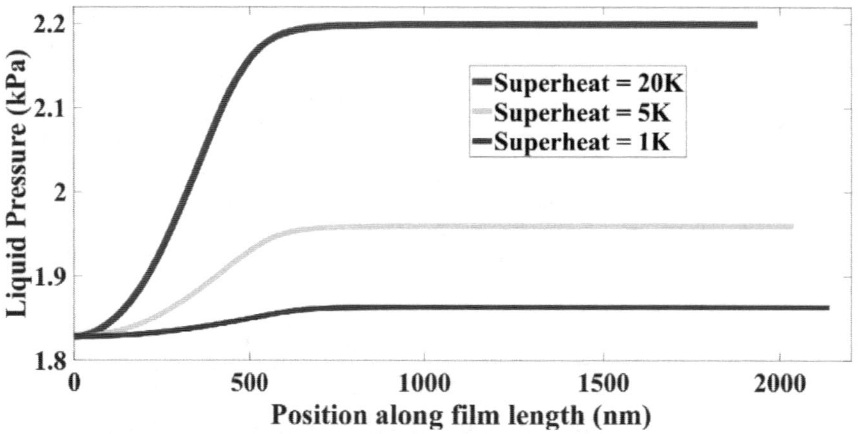

Figure 6. Effect of superheat on the liquid pressure

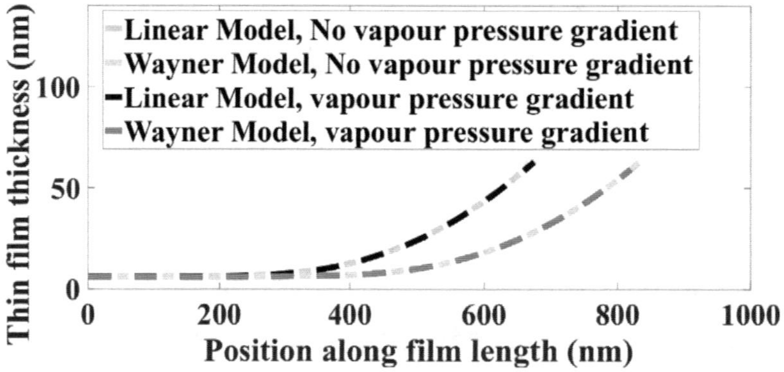

Figure 7. Effect of superheat on the liquid pressure

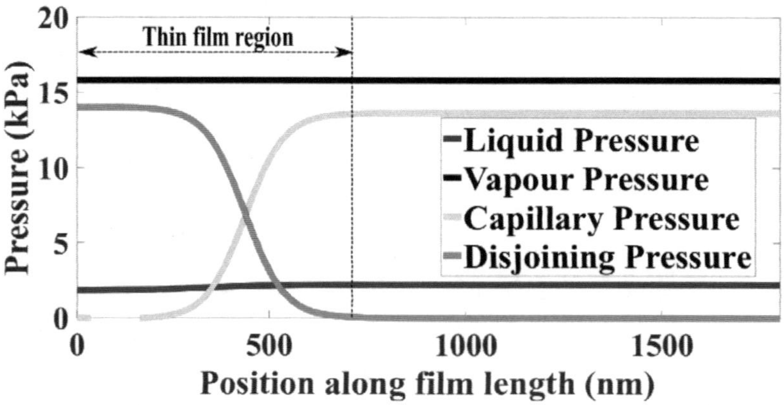

Figure 8. Variation of different pressure components along the meniscus length

IOP Conf. Series: Materials Science and Engineering 1139 (2021) 012013 doi:10.1088/1757-899X/1139/1/012013

$$P_l = P_v - \frac{\bar{A}}{\delta^3} - \sigma \frac{\frac{d^2\delta}{dx^2}}{\left\{1 + \left(\frac{d\delta}{dx}\right)^2\right\}^{3/2}}$$

Table 1 shows different values of first and second derivative of film thickness available in literature. As can be seen from the table their values are taken from Akkus and Dursunkaya [10] and Park et al [13], although Du and Zhao [9] have advised not to use the same. They are still used because it is relatively easy to implement them as compared to far field boundary conditions.

The mass flow rate of vapour should always be zero at the end of the thin film region [13,19] and the mass flow rate of liquid is found out by ignoring the vapour pressure gradient while running the calculation. Whatever the liquid flow rate that we get is substituted into the vapour flow rate at x = 0.

δ_0 is calculated by any of the following two ways:

- **Using Wayner's mass flux model:** Here, $\dot{m}'' = 0$ and $T_{lv} = T_w$ are substituted in equation (18) (At $x = 0$, there is no heat transfer, so, the mass flux is taken as zero and since the film is very thin here, the interface temperature is almost the same as the wall temperature). Thus, the following expression is derived:

$$\delta_0 = \left\{\frac{\bar{A}T_v}{\rho_l \lambda (T_w - T_v)}\right\}^{1/3} \tag{22}$$

In this study, this method will be termed Wayner Method.

- **Using Clausius-Clapeyron equation:** Here, the values of $\dot{m}''$ and T_{lv} at $x=0$ are substituted in Clausius–Clapeyron equation and the following expression is derived:

$$\delta_0 = \left\{\frac{\bar{A}T_v}{\rho_v \lambda (T_w - T_v)}\right\}^{1/3} \tag{23}$$

This method will be termed Clausius–Clapeyron Method.

As can be seen from the system of boundary conditions, proper calculation of δ_0 is crucial for determination of liquid pressure at x = 0 and thus for overall solution. Figure 3 compares the various parameters at x = 0 calculated from the two methods. Working fluid considered are some of the common fluid used in these problems i.e octane, ammonia, acetone and pentane. The saturation temperature was kept at a constant value of 30°C and superheat of 1K. From figure 3(a), it is clear that δ_0 calculated from equation (22) is very low. This gives negative liquid pressure (figure 3(b)) which does not satisfy the underlying physics of the problem. The same problem was also reported by Wang et al. [20]. Although various researchers ([3], [4]) have advised to take a slightly higher value than calculated, the problem still persist. Equation (23) on the other hand, gives high value of δ_0. This gives relatively thick film at x = 0 which leads to low value of disjoining pressure (figure 3(c)). This also goes against the nature of the problem. Simplest solution is to take a value in between the values calculated from equation (22) and (23). In this paper, an attempt is made to define $(\delta_0)_{min}$ and $(\delta_0)_{max}$ which would satisfy the physics of thin films. As can be seen from figure 2, x = 0 point lies in between the adsorbed region and the thin film region. As such at that point evaporation has not started yet and disjoining pressure still dominates. Thus, the governing equation here is a slightly modified form of equation (2).

$$P_v - P_l = P_d$$

$$\Rightarrow P_v - P_l = \frac{\bar{A}}{\delta_0^3} \tag{24}$$

Now, as $\delta_0 \rightarrow (\delta_0)_{min}$, $(P_v - P_l) \rightarrow P_v$. So, equation (24) becomes:

$$P_v = \frac{\bar{A}}{(\delta_0)^3_{min}}$$

$$(\delta_0)_{min} = \left(\frac{\bar{A}}{P_v}\right)^{1/3} \tag{25}$$

$(\delta_0)_{max}$ is assigned the value calculated from equation (23). Next, average or any value between $(\delta_0)_{min}$ and $(\delta_0)_{max}$ is taken to be δ_0.

3. Results

The system of equations is solved taking octane as working fluid and saturation temperature, 70°C. To verify the results, mass flow rate of vapour and thin film profile are plotted for different superheat. As stated earlier, mass flow rate of vapour is always zero at the end of the thin film region. Figure 4 clearly exhibit such a trend. Figure 5 shows that the thin film thickness increases with increase in superheat (both overall and at the onset of thin-film region) but its length decreases which agrees with the results found in the literature. The reason is that high superheat increases the rate of evaporation and thus more liquid is pumped into the the thin film. This also increases the liquid pressure as is clear from figure 6.

Figure 7 shows how the vapour pressure gradient affects the profile of thin film for different evaporating mass flux model. As can be clearly seen from the figure, vapour pressure gradient plays a negligible role in the thin film region for capillary cooling devices. This can also be confirmed from Figure 8 where effect on different components of pressure (capillary, disjoining, liquid and vapour) along the film length is shown. The figure clearly shows that the disjoining pressure decreases and the capillary pressure increases as the film progresses from the thin film region to the capillary region as expected. However, the vapour pressure remains constant throughout the extended meniscus. As such the vapour pressure gradient can be neglected. This justifies the assumption of constant vapour pressure by most of the researchers.

4. Conclusion

Following are the conclusions:

i. δ_0 calculated from both equations (22) and (23) do not satisfy the physics of the problem and may give unrealistic solution.
ii. The vapour pressure is uniform throughout the extended meniscus of two phase capillary devices and hence its gradient can be neglected.
iii. A proper choice of the evaporating mass flux model is critical in the analysis of evaporating thin films of two-phase capillary devices, which is very important for predicting their heat transfer characteristics.

Nomenclature

General Symbols

$\bar{A}$	Dispersion constant (J)
$\bar{M}$	Molecular mass (kg/kmol)
$\bar{R}$	Universal gas constant (J/kmol-K)
H	Channel height (m)
k	Thermal conductivity (W/mK)
L	Length (m)
$\dot{m}''$	Interface net mass flux (kg/m^{2}s)
$\dot{m}$	Mass flow rate (kg/ms)
P	Pressure (Pa)
q''	Heat flux (W/m^2)
T	Temperature (K)

Greek symbols

β	Slip coefficient
δ	Thin film thickness (m)
λ	Latent heat of vaporisation (J/kg)
μ	Coefficient of viscosity (Pa-s)
$\hat{\sigma}$	Accommodation coefficient
τ	Shear stress (N/m^2)

Subscripts

c	Capillary
cc	Clausius-Clapeyron
d	Disjoining
l	Liquid
lv	Liquid-vapour interface
max	Maximum
min	Minimum
sat	Saturation
v	Vapour
w	Wall
way	Wayner

Abbreviations

FoM	Figure of merit

ρ	Density (kg/m^3)	HP	Heat pipe
σ	Surface tension (N/m)	IL	Interline

References

[1] Wayner Jr P and Coccio C 1971 *AIChE journal* **17** 569–574
[2] Wayner P, Kao Y and LaCroix L 1976 *International Journal of Heat and Mass Transfer* **19** 487–492
[3] Wang H, Garimella S V and Murthy J Y 2007 *International Journal of Heat and Mass Transfer* **50** 3933–3942
[4] Hanchak M S, Vangsness M D, Byrd L W and Ervin J S 2014 *International Journal of Heat and Mass Transfer* **75** 196–206
[5] Jasvanth V, Ambirajan A, Kumar D and Arakeri J H 2013 *Journal of Thermophysics and Heat Transfer* **27** 633–640
[6] Wayner Jr P 1978 *Journal of Heat Transfer* **100** 155–159
[7] Wayner Jr P 1978 *International Journal of Heat and Mass Transfer* **21** 362–364
[8] Wayner P C 1989 *Journal of Heat Transfer* **111** 813–815
[9] Du S Y and Zhao Y H 2011 *International Journal of Heat and Mass Transfer* **54** 3694–3701
[10] Akkuş Y and Dursunkaya Z 2016 *International Journal of Heat and Mass Transfer* **101** 742–748
[11] Akkuş Y, Tarman H I, Çetin B and Dursunkaya Z 2017 *International Journal of Thermal Sciences* **121** 237–248
[12] Fu B, Zhao N, Tian B, Corey W and Ma H 2018 *Journal of Heat Transfer* **140** 011502
[13] Park K, Noh K J and Lee K S 2003 *International Journal of Heat and Mass Transfer* **46** 2381–2388
[14] Kou Z H, Lv H T, Zeng W, Bai M L and Lv J Z 2015 *International Communications in Heat and Mass Transfer* **63** 49–53
[15] Ahmed S and Pandey M 2019 *Physics of Fluids* **31** 092001
[16] Ahmed S and Pandey M 2020 *International Communications in Heat and Mass Transfer* **117** 104803
[17] Moosman S and Homsy G 1980 *Journal of Colloid and Interface Science* **73** 212–223
[18] Stephan P C and Busse C A 1992 *International Journal of Heat and Mass Transfer* **35** 383–391
[19] Zhao J J, Duan Y Y, Wang X D and Wang B X 2011 *Journal of Nanoparticle Research* **13** 5033–5047
[20] Wang H, Pan Z and Chen Z 2009 *Frontiers of Energy and Power Engineering in China* **3** 141–151

IOP Conf. Series: Materials Science and Engineering 1139 (2021) 012004 doi:10.1088/1757-899X/1139/1/012004

A method of fast quality control inspection of loop heat pipe

Bin-Juine Huang[1,*] , H H Huang[1], T L Liang[1], P E Yang[1], S S Hu[1]

[1]Department of Mechanical Engineering, National Taiwan University, Taipei, Taiwan

Abstract. The dynamic behaviour of loop heat pipe (LHP) changes with heat load and physical properties of LHP such as fluid charging, wick material, and tolerance in manufacture. The start-up phenomena of a LHP can be classified into four modes, according to the heat load Q_{in}: failure mode , oscillating mode, overshoot mode, and normal mode . It is found that the operation mode is related to the rate of temperature rise in evaporator during initial transient performance using a step heating load. The abnormal high rate of temperature rise can be observed for LHP with failure, oscillating or overshoot mode performance. Based on this phenomenon, the normal-mode performance of LHP can be assured from a short transient test in about 80s using a step heat load, instead of using steady-state test (taking 1,800s). This transient test method has been applied in the quality inspection of commercial loop heat pipe in mass production. No LHP failure was found for more than 10 years for about 30,000 sets of street LED luminaires designed with LHP.

1. Introduction

Loop heat pipe (LHP) is a passive heat-transfer device that uses micro-scale wick technology to transfer a large amount of heat to a distant place through a connecting pipe. There is a startup condition for a LHP [1-3] that requires a temperature difference across the wick structure.

The phenomena of minimal start-up heat load and temperature overshoot was observed. Start-up of LHP is a behaviour of dynamic system that is very complex. The factors affecting the start-up of LHP include working fluid used, design of compensation chamber and evaporator, the initial state of working fluid in the evaporator, the charge volume of working fluid, LHP attitude, the initial state of LHP before heat load applied, etc. Two phenomena were observed very often during the start-up of LHP: (1) temperature overshoot; and (2) minimal heat load required to start up.

The quality inspection test of LHP may be carried out by measuring the steady-state performance. But it will take a long time for reaching a steady state, at least 1800s (30 min), and the cost is very high.

The transient behaviour of LHP can be used to distinguish the normal or abnormal operation of LHP in steady state. The present study tries to develop a fast transient test method for LHP. A fast inspection method in quality control of a mass production is then developed according to the transient behaviour of LHP.

2. System dynamics behaviour of LHP

The system dynamic behaviour of LHP can be studied by building a system dynamics model [4]. The start-up phenomena of a LHP can be classified into four modes, according to the heat load Q_{in}: (1) failure

* corresponding author: bjhuang@seed.net.tw

mode when $Q_{in} < Q_{min}$, (2) oscillating mode when $Q_{min} < Q_{in} < Q_{crit}$, (3) overshoot mode when $Q_{crit} < Q_{in} < Q_s$, and (4) normal mode when $Q_s < Q_{in}$.

The minimum heat load Q_{min} is defined as the condition that a LHP can start up successfully. If the heat load Q_{in} is lower than Q_{min}, start-up will fail (failure mode).

When the heat load Q_{in} is lower than Q_{crit}, temperature oscillation occurs (oscillating mode). The vapor flow to the condenser is in oscillating or pulsating state. As the heat load Q_{in} is higher than Q_{crit}, LHP enters the overshoot mode in start-up and an overshoot of temperature will be observed.

When the heat load Q_{in} is higher than Q_s, a smooth start-up will occur. Table 1 shows the design specification of the LHP used in the present study (Figure 1).

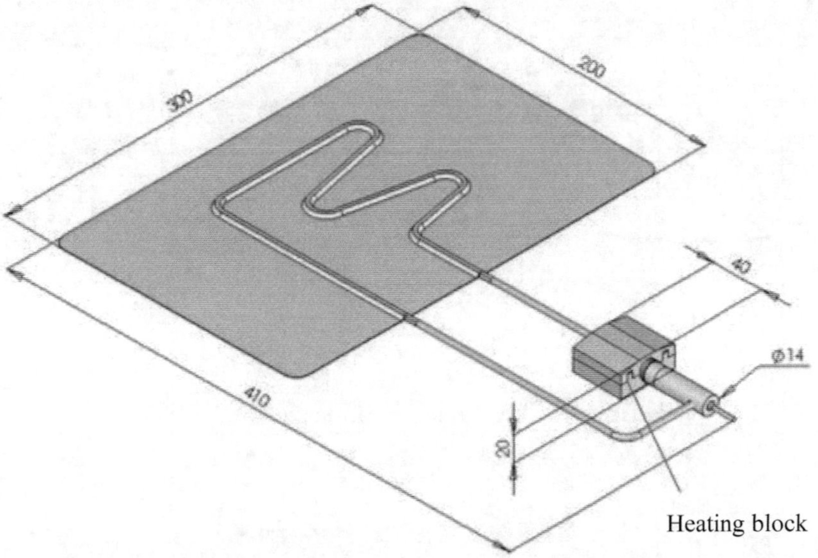

Figure 1. Design of the LHP for experiment.

Table 1. Specification of the LHP used in the study.

Evaporator:	capillary force	50 cm-Hg
	porosity	60%
	pore size	5-10 μm
	outer diameter	14 mm
	total length	90 mm
Connecting Tube:	material	OFCU
	outer diameter	4 mm
	inner diameter	2.3 mm
Condenser Plate:	material	OFCU
	dimension	200mmx 300m x 0.5mm
Heating Block:	material	Aluminum alloy 6063
	dimension	45mm x 40mm x 20mm
Working Fluid:	Acetone	

Experimental analysis of the LHP transient performance at different operating conditions was carried out to study the characteristic operating modes of LHP [4].

For $Q_{in} <$ 5W (Q_{min}), start-up failure is observed (Figure 2). For 20W(Q_{crit}) $< Q_{in} <$40W(Q_s), oscillation phenomenon is observed (Figure 3). For heat load $Q_{in} <$ 40W(Q_s), overshoot phenomenon is observed (Figure 4). When the heat load $Q_{in}>$40W(Q_s), a smooth start-up is observed (Figure 5). A smooth transient temperature rising curves are observed at condenser inlet. This is supposed to be the normal-mode operation of a LHP.

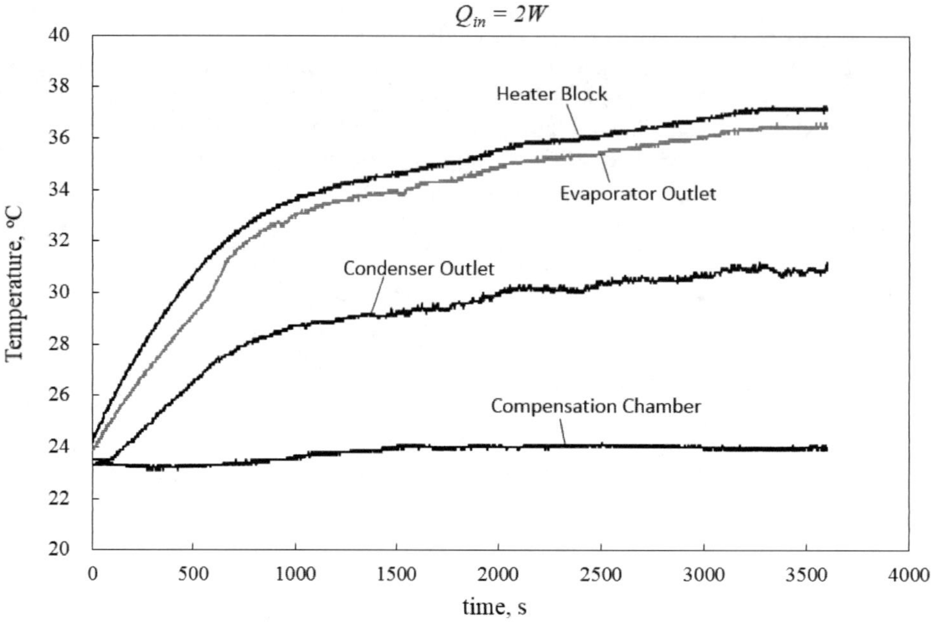

Figure 2. Failure mode performance (Q_{in} = 2W).

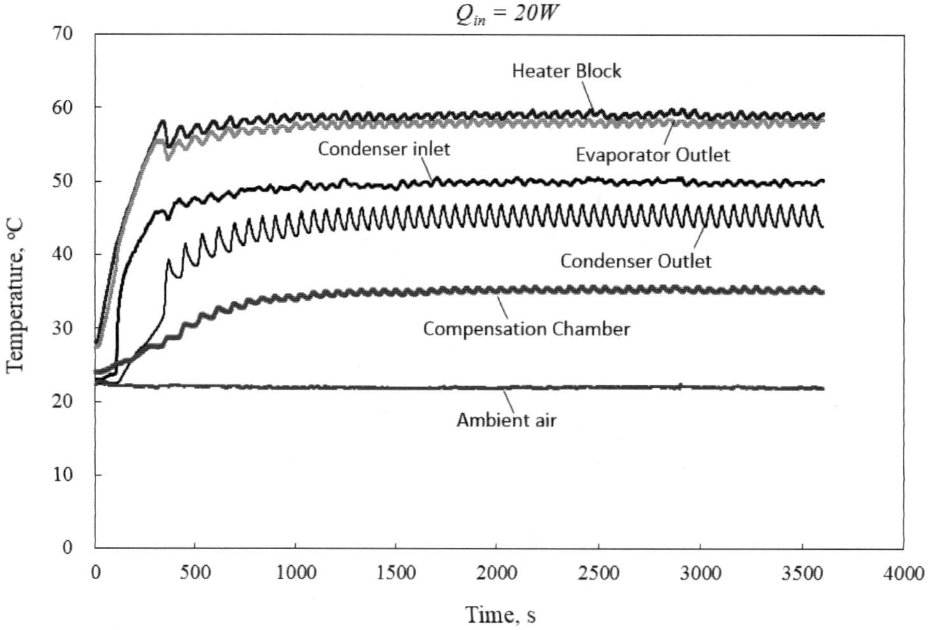

Figure 3. Oscillating mode performance (Q_{in} = 20W).

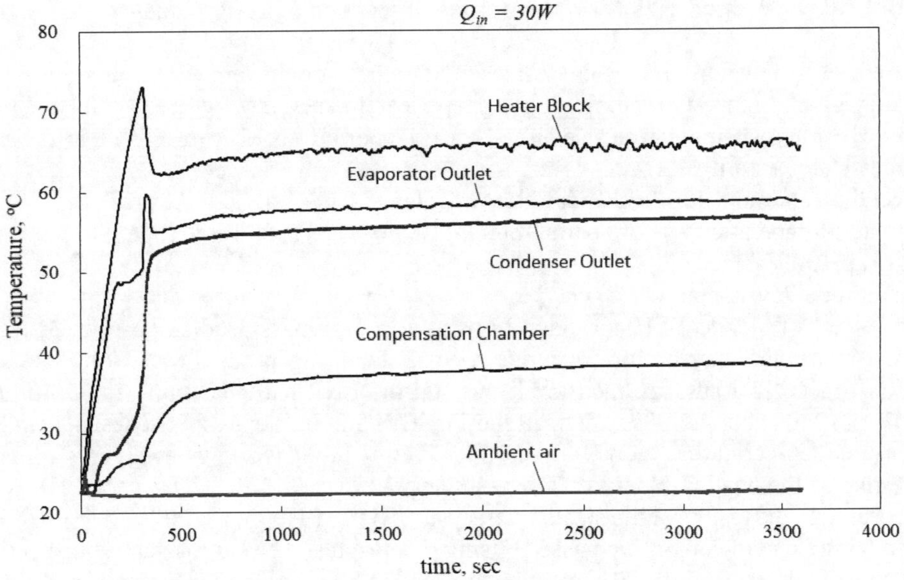

Figure 4. Overshoot mode performance (Q_{in} = 30W).

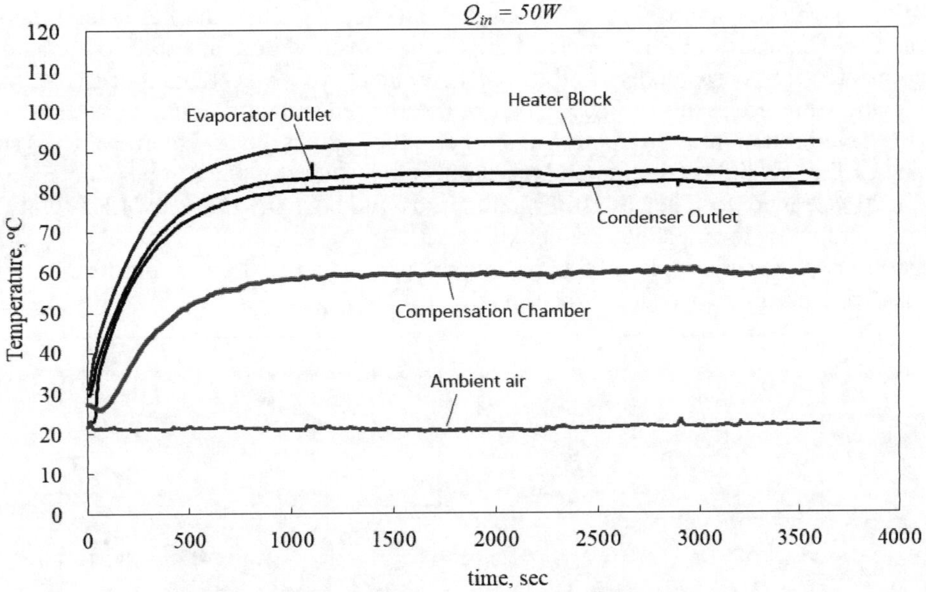

Figure 5. Normal mode performance (Q_{in} > 30W).

3. Fast inspection in quality control of LHP

The quality inspection test of LHP may be carried out by measuring the steady-state performance of each LHP. But it will take a long time for reaching a steady state, at least 30 minutes (1800s), and the cost is very high.

It can be seen from Figure 2 to Figure 5 that, the rate of the heating flock temperature rise at the start-up period is quite different at different heat load Q_{in}. It is observed that the temperature rise rate per heating load (K $s^{-1}W^{-1}$) is much higher if LHP does not operate normally. This dynamic or transient behaviour of LHP can be used as the criterion for fast inspection of LHP in quality control during mass production.

The present study develops a transient test method to measure the temperature rise, $\triangle T_b = T_b(t) - T_{bi}$, of the heating block right after employing a step heat load 100W where T_{bi} is the initial block temperature. The rate of temperature rise at a short time period may be used to identify the operation mode at steady state performance.

We select the conditions for the standard test as folows:

• Key parameter: temperature rise of heating block $\triangle T_{bo} = T_b(t_o) - T_{bi}$ at fixed time interval t_o
• Step heat load: 100W (fixed)
• Operation attitude: horizontal (0°)

We use 11 sets of identical LHPs randomly picked from a mass production line [5] for the study. Their thermal resistance may be different due to manufacturing process variation. The steady-state performance of these 11 LHPs are measured first and the LHP with start-up failure are identified as LHP02, LHP03, LHP07 and LHP09. Then, all the 11 LHPs are run step response test at heat load 100W.

The measured temperature response curve of the heating block is shown in Figure 6. It is found that temperature rise of the heating block ΔT_{bo} at t_o= 80 s will exceed 30°C (ΔT_{crit}) for LHP with start-up failure. According to the results of the steady-state test, LHP02, LHP03, LHP07 and LHP09 are the LHP with start-up failure and the others are passed. Using this fast test method for quality inspection, we can shorten the time of the test from 1,800 s (steady-state test) to 80 s. Figure 7 is the transient test equipment of LHPs developed for the transient test of six LHPs simultaneously.

This test method can be applied to any other LHP with different design or manufacturing process. The calibration test (steady-state test) must be carried out first using the samples taken randomly from a production line. The test results of sampled LHPs with start-up failure are used to compare with the transient temperature response curves of all the LHPs to identify GO or NO-GO of LHPs, from which the critical temperature rise at time t_o , ΔT_{crit} , can be determined and used in transient test.

The LHPs have been in mass production and used in the heat dissipation of 100W LED street light. One thousand LED street lights (commercial product) were installed on the campus of National Taiwan University in 2008 (Figure 8). They are still running today without any failure of LHP. Only failure of LED power adaptors was found, about 2% per year.

The measurement uncertainty is ±0.5^{o}C in temperature T_b and ±10% in input heating power Q_{in} . This error has been taken into account in the selection of $\triangle T_{crit}$.

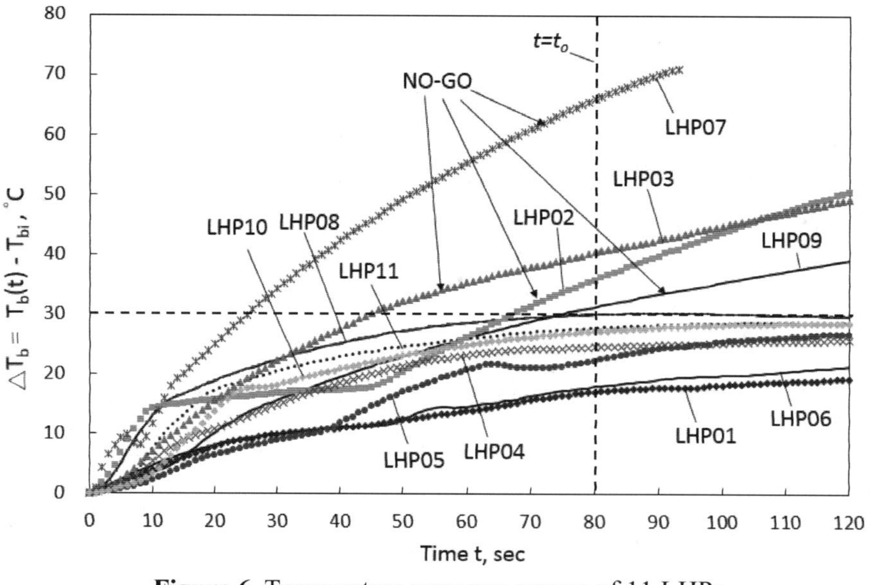

Figure 6. Temperature response curves of 11 LHPs.

Figure 7. Transient test equipment of LHPs.

Figure 8. 1000 LED street lights using LHP installed on campus of NTU (2008).

4. Conclusions

The present study develops a transient test method for fast inspection in the commercial production of LHP. The rate of temperature rise (T_b-T_{bi}) of the LHP heating block right after employing a step heat load is used to determine the normal or abnormal performance of LHP. Using this fast test method, we can reduce the time of the test from 1,800 s (steady-state test) to 80 s. For the 100W LHP production, the LHP is identified as abnormal (NO-GO) if ΔT_{bo} is higher than ΔT_{crit} (chosen as 30°C) at t_o=80 s. This fast inspection method in quality control has been used in the mass production of LHP [5]. No LHP failure occur for about 30,000 sets installed in LED luminaires during the past 10 years.

This test method can be applied to any other LHP with different design or manufacturing process. The calibration test (steady-state test) must be carried out first. The steady-state test results of sampled LHPs with start-up failure are used to identify the conditions for LHPs with start-up failure, i.e. finding the critical temperature rise at time t_o , ΔT_{crit} . By selecting a proper critical temperature rise at time t_o , ΔT_{crit} , LHPs can be identified as GO or NO-GO in the transient test and the QC test can speed up.

5. References

[1] Maidanik Yu F Loop Heat Pipe Technology, Institute of Thermal Physics, Ural Branch of the Russian Academy of Sciences, Ekaterinburg, Russia.
[2] Maidanik Yu F and Pastukhov V G 1999 Loop Heat Pipe-Recent Developments, Test Results and Applications, Institute of Thermophysics, Ural Division of Russian Academy of Sciences, SAE Technical Paper 1990-01-2530
[3] Maydanik Y 2005 Review of loop heat pipes, *Appl. Therm. Eng.* **25** (5&6) 635–657
[4] Huang B J, Huang H H, Lian T L 2009 System dynamics model and startup behavior of loop

heat pipe. *Appl. Therm. Eng.* **29** 2999-3005
[5] Huang B J, Chuang H Y, Yang P E 2017 Low-cost manufacturing of loop heat pipe for commercial applications. *Appl. Therm. Eng.* **126**, 1091-1097.

Acknowledgement
This study was supported by Advanced Thermal Devices (ATD) Inc (Taiwan) and Ministry of Science and Technology, Taiwan, through Grant: NSC93-2212-E-002-032.

IOP Conf. Series: Materials Science and Engineering 1139 (2021) 012011 doi:10.1088/1757-899X/1139/1/012011

Loop heat pipe dynamics modeling and analysis

Vasily Buz[1*] and Konstantin Goncharov[1]

[1]TAIS Ltd, Khimky, Moscow region, Russia

Abstract. The report presents mathematical model and computational analysis of dynamic processes in the loop heat pipe, as well as some ways of regulating its work. This is considered a one-dimensional unsteady model of the loop. For each element of loop the model contains the energy equation for flow of fluid and to the walls, and the equation of conservation of momentum of the working fluid. The joint solution of a system of such equations for all elements of the loop allows to calculate the temperature distribution and pressure contour at any point in time. The calculations showed that under certain conditions in the loop heat pipe can be auto-oscillations of different nature. Analysed methods of eliminating these auto-oscillations. One way to regulate the loop heat pipe is the use of a control valve on the additional bypass line. The analysis showed that this method can significantly improve the quality of temperature control and reduce energy consumption by regulation or do without energy consumption. Presents a model of the dynamics of loop heat pipe gives physically reasonable results of calculations and is consistent with the results of the tests.

1. Introduction

One of the most effective devices for diverting heat from both ground and on-board equipment is loop heat pipes (LHP) [1]. In some cases, it is necessary to ensure temperature stability when the working conditions of the equipment change. To do this, additional measures should be taken to regulate the contoured heat pipe. One such measure may be the use of a bypass line with a valve-regulator, which ensures the transfer of part of the vapor from the vapor line to the compensation chamber. This leads to the displacement of the liquid from the compensatory cavity into the capacitor, the reduction of the condensation zone and the increase in the temperature of the evaporator. By regulating the consumption of vapor in the bypass line with the help of a valve-regulator can regulate the temperature of the vaporizer. In practice, this method has shown its performance and reliability [2].

For successful design of LHP-based thermo-stabilization systems, calculation LHP models with bypass and valve-regulator in its composition are needed. The development of the LHP mathematical model has been carried out by different authors for a long time [3-11], but even in a stationary form the model has problems due to complex and interconnected heat-hydraulic processes in LHP.

The dynamics model of LHP without the bypass line, is also of independent importance. It allows you to calculate the launch of LHP, transients, the occurrence of auto-oscillations, and identify possible problems of stable LHP operation. This is important because LHP is an autonomous self-regulating device.

* corresponding author: shatuny@yandex.ru

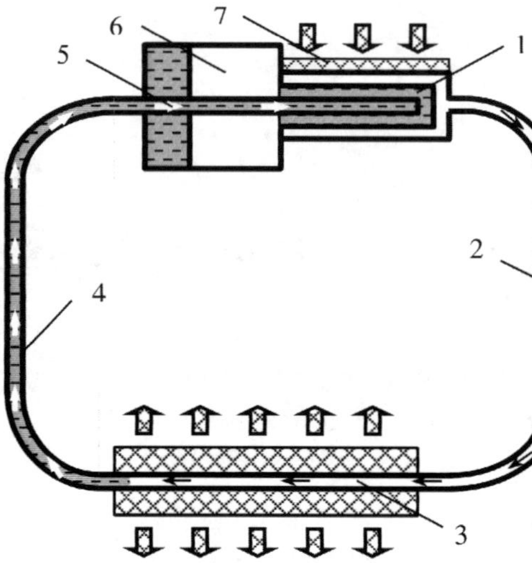

Figure 1. Schematics of the simplified circuit LHP: 1 - porous wick; 2 - vapor line; 3 - condenser; 4 - condenser line; 5 - segment condenser line inside the reservoir; 6 - reservoir; 7 - pump module head; 8 - bypass line; 9 - bellows; 11 - valve.

2. Modeling of the simplest LHP

2.1. Mathematical modeling

An important feature of loop heat pipes is the self-regulation of their work. It is determined by a complex of interrelated heat and hydraulic processes. Numerous internal feedbacks, the physical nature of which is not always clear, make it difficult to use traditional universal methods of calculation.

As a first step we will consider model of LHP dynamics having simplified circuit (Figure 1). The loop has been divided into the j elements shown in Figure 1 and for each j-th element of the loop we will write down the equations of energy for the working fluid flow and for a wall, and also the conservation of momentum equation of the working fluid:

$$\frac{\partial(\rho_f h_f)}{\partial\tau}+\frac{\partial(\rho_f u_f h_g)}{\partial z}=\lambda_f\frac{\partial^2 t_f}{\partial z^2}+q_f \quad (1)$$

$$\rho_w c_w\frac{\partial t_w}{\partial\tau}=\lambda_w\frac{\partial^2 t_w}{\partial z^2}+q_w \quad (2)$$

$$\frac{\partial p_f}{\partial z}=\frac{\xi}{d}\frac{\rho_f u_f^2}{2}+G\frac{\partial u_f}{\partial z} \quad (3)$$

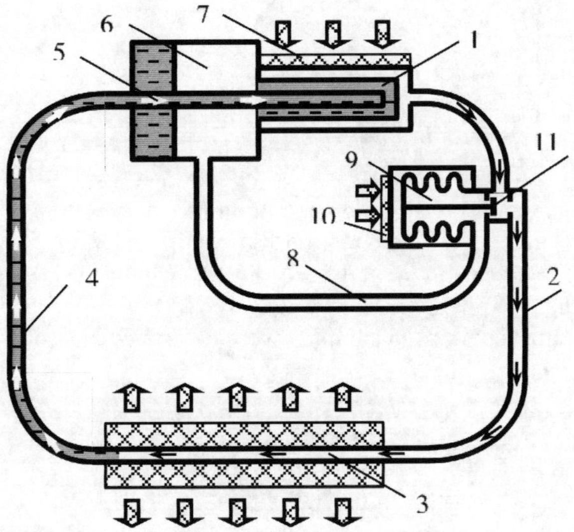

Figure 2. Schematics of the simplified circuit LHP with a valve-regulator: 1 - porous wick; 2 - vapour line; 3 - condenser; 4 - condenser line; 5 - segment condenser line inside the reservoir; 6 - reservoir; 7 - pump module head; 8 - bypass line; 9 - bellows; 11 - valve.

Boundary conditions for these equations (1)-(3) we will write down as:

At z=0
$$\begin{cases} (h_f)_j = (h_f)_{j-1} \\ (t_w)_j = (t_w)_{j-1} \\ (p_f)_j = (p_f)_{j-1} \end{cases} \tag{4}$$

At z=l$_j$
$$\begin{cases} \left(\frac{\partial h_f}{\partial z}\right)_j = \left(\frac{\partial h_f}{\partial z}\right)_{j-1} \\ \left(\frac{\partial t_w}{\partial z}\right)_j = \left(\frac{\partial t_w}{\partial z}\right)_{j-1} \end{cases} \tag{5}$$

For the vapour line at z=0 it is accepted

$$(p_f)_1 = (p_f)_4 + \Delta p \tag{6}$$

Where $\Delta p \le \frac{2\sigma}{R}$ is the general pressure losses in a LHP. For the porous cartridge at z=l$_j$ it is accepted that:

$$\left(\frac{\partial t_w}{\partial z}\right)_4 = \left(\frac{\partial t_w}{\partial z}\right)_1 + \frac{Q}{\lambda_{w,j} S} \tag{7}$$

The correlation linking the enthalpy and the temperature of the working fluid was determined with the help of the following rationale:

$$\begin{cases} if\ h' = h'_s & t = t_s + \frac{h'_s - h}{c\prime} & x = 0 \\ if\ h'_s < h < h" & t = t_s & x = \frac{h - h'_s}{r} \\ if\ h' > h"_s & t = t_s + \frac{h - h"_s}{c"} & x = 1 \end{cases} \tag{8}$$

Where x is the vapour quality. Saturation temperatures of working fluids in each section were determined as function of pressure on the equations describing relation of pressure and temperature on the saturation line. The homogeneous model according to which the density was determined on the equation was applied to the two-phase working fluid.

Source components in the equations (1) and (2) were determined as follows:

$$q_f = \frac{\alpha_{in}(t_w - t_f)b_{in}}{\omega_f} \tag{9}$$

$$q_w = \frac{[\alpha_{ext}(t_{ext} - t_w)b_{ext} - \alpha_{in}(t_w - t_f)b_{in} + q_l]}{\omega_w} \tag{10}$$

With help of the combined equation of balance of heat and weights for the reservoir

$$r\left(\frac{dm"_{fR}}{d\tau} - G_4 x_4\right) = k_R(t_1 - t_R) + (\alpha_{ext})_R(t_R - t_4)S_4 - C_R\frac{dt_R}{d\tau} - Gc'_4(t_R - t_4) \tag{11}$$

it is possible to determine weight of the vapour of working fluid in the reservoir at any moment of time. The working fluid's vapour density in the reservoir was determined on the ratio

$$(\rho_f)_R = \frac{m"_{f,R}}{V"_R} \tag{12}$$

Where volume of a vapour phase in the reservoir,

$$V"_R = \frac{V_{R\Sigma}\rho' - [(m_f)_0 - (m_f)_\Sigma]}{\rho' - \rho"} \tag{13}$$

and $(m_f)_\Sigma$ is the weight of the working fluid in the loop expressed by:

$$(m_f)_\Sigma = \Sigma \int_0^l (\rho_f \omega_f \partial z)_i + (m')_R + (m")_R \tag{14}$$

Where the temperature t_R of working fluid in the reservoir was determined during solution of the written down system of equations. Density of the vapour calculated on equation (12) is equal to density of saturated vapour at the temperature t_R.

The temperature t_7 of temperature-stabilized item at any moment could be determined from the equation of thermal balance:

$$Q_7 - C_7\frac{dt_7}{d\tau} - \alpha_{ext}S(t_7 - t_{ext}) = k_7(t_7 - t_1) \tag{15}$$

2.2. Modeling of LHP temperature regimes when changing heat load

The results of loading step change calculations (5 W and 15 W) are presented in Figure 3. As it can be seen, this change at absence of any measures on regulation corresponds to change 4.5° C of the temperature-stabilized item's temperature.

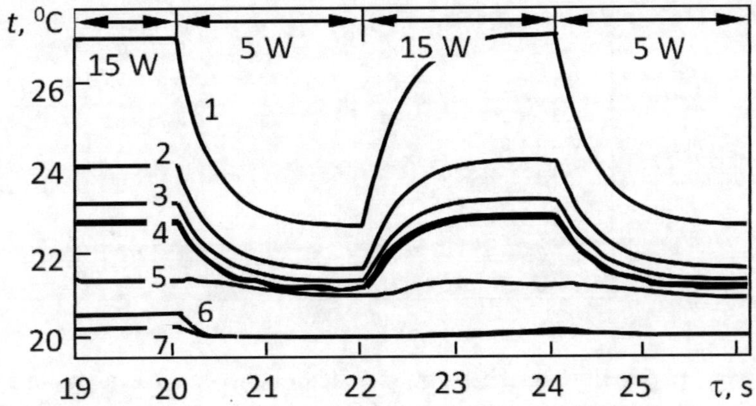

Figure 3. Behavior in time of the temperature in various points of the LHP at step change of the power. (1 – pump module head, 2 - evaporator wall, 3 - saturation in the evaporator, 4 - saturation in the reservoir, 7 - output from the condenser).

The temperature difference between the cooling object and the environment varies from 7 to 2.5 degrees. In time there is a redistribution of the working fluid rate. At that the some periodic filling and draining of the condenser are occurred.

The results of numeric analysis of the LHP startup and transition from one mode to another. It is shown that these processes are not always successful (curves 1-3 in Fig. 3). The key parameters that determine their success are the mass of the working fluid charged and a heat capacity of compensation chamber.

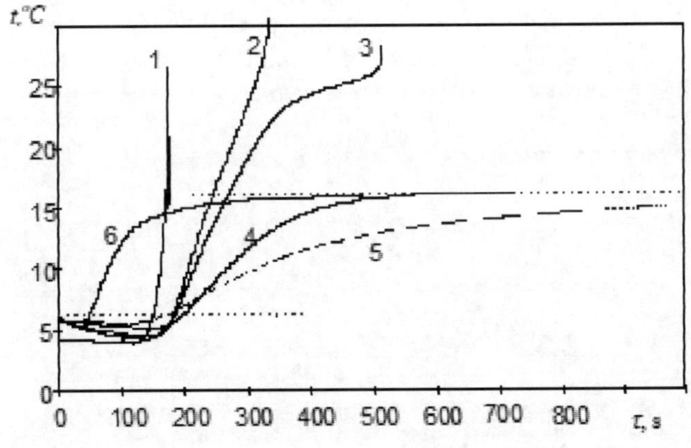

Figure 4. The LHP evaporator time dependencies when control is absent and heat removal zone effective temperature is suddenly changed from -45°C to -37°C when different values of compensation chamber total heat capacity (mc , J/K): 1- 0; 2 - 10; 3- 20; 4 - 50; 5 - 100 and when d4 = 1 mm, mc = 20 J/K (curve 6). Shade line - are stationary regimes temperatures (-45°C and -37°C).

2.3. Auto-oscillation at LHP start-up

Results of calculations LHP characteristics with length of a vapor line and liquid line 0.2 m, the condenser is 0.5 m, diameter of a vapor line and the condenser is 1 mm, liquid line is 0.6 mm are presented on Figure 5.

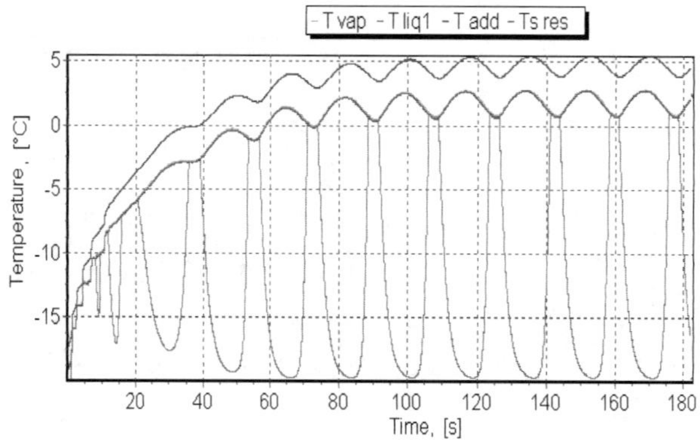

Figure 5. Start-up of LHP when a heat transfer factor between the evaporator and reservoir negligible: 1 – temperature of the evaporator; 2 – saturation temperature; 3 – temperature on an outlet of condenser.

The working fluid is ammonia. Evaluation of the evaporator over the condenser is absent. Cooling of the condenser is convective. A heat load of evaporator is 30 W.

Calculation results is shown, that at small value of a heat transfer factor between evaporator and reservoir, there can be significant self-oscillations of the liquid, accompanied moving of vapor from the condenser to reservoir and return liquid flow from reservoir. At increase in a heat transfer between the evaporator and reservoir auto-oscillations decrease and become insignificant.

2.4. Influence of the variable evaporator heat load on the quality of thermal stabilization

An example of the calculation of LHP for a stepwise change in the heat load on the evaporator is considered. Figure 6 shows the graph of temperature variation in the loop without a regulator. If the load varies from 50 to 200 W, the temperature of the evaporator (line 1) can be changing by 15 degrees. Moreover, thermal-hydraulic self-oscillations of temperature are observed because of the "breakthrough" of steam from the condenser to the reservoir.

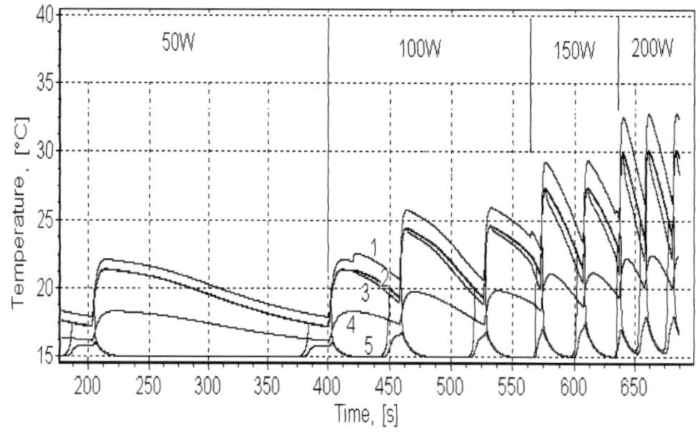

Figure 6. The calculated time variation of temperature in different points of the LHP without the regulators when changing the heat load of the evaporator. 1 – evaporator temperature, 2 – reservoir temperature, 5 – outlet temperature from the condenser

Figure 7 shows the results of the LHP tests for different values of the heat load on the evaporator. As can be seen, fluctuations in the temperature of the evaporator are observed, the amplitude of which, as in the calculations, is of the order of 5 degrees.

An important parameter of the temperature control system is the quality of thermo-stabilization. In LHP to improve the quality of temperature stabilization uses additional units - device of heat action, hydraulic regulators with bypass lines or without them. Conducted design analysis of the LHP work with such devices (Figure 8). Conditions of quality temperature stabilization is defined.

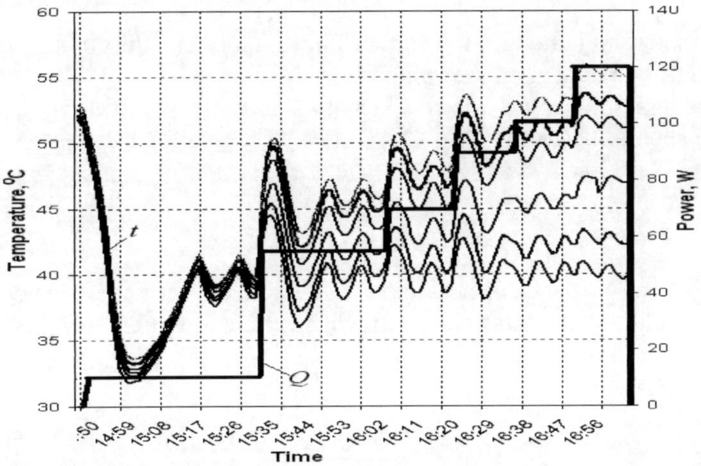

Figure 7. LHP test results. Working fluid - ammonia, condenser cooling temperature - 150C, evaporator heat power - 0-120 W.

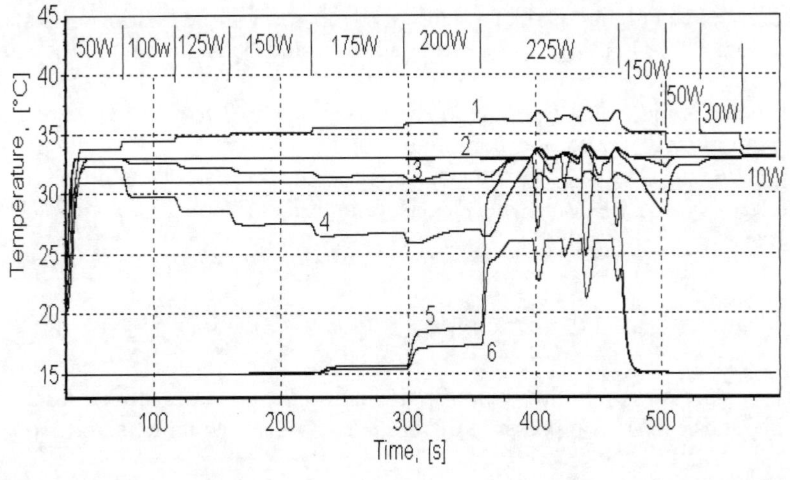

Figure 8. The calculated time variation of temperature in different points of the LHP with regulators when changing the heat load of the evaporator. 1 – evaporator temperature, 2 – reservoir temperature, 5 – outlet temperature from the condenser.

3. Modelling of the LHP with bypass regulation

Circuit of the LHP with bypass and a valve - regulator is presented on Figure 2. As a first approximation by processes of heat exchange in bypass lines it is neglected. We suppose that hydraulic resistance bypass is not enough pipeline in comparison with resistance of the valve.

The working fluid rate through bypass line G_8 is determined, on the one hand, by difference of pressure between vapour line 2 and the reservoir 6, on the other hand - factor of hydraulic resistance of the valve 11. If we assume that inertia forces have negligible role, the rate could be found from following equation

$$p_2 - p_6 = \zeta_{V8}\frac{G_8^2}{2\rho''\omega_8^2} - \frac{G_2^2}{2\rho''\omega_2^2} \tag{16}$$

In case of LHP with bypass line the design model presented in section 1.1, will contain the following features:

1) The working fluid rate through vapour line and further - the condenser and condenser line should be reduced on the value;
2) Pressure losses in vapour line will increases on value of valve–regulator's local losses that are approximately equal to:

$$\Delta p = \zeta_{V2}\frac{G_2^2}{2\rho''\omega_2^2} \tag{17}$$

3) The equation of balance of heat and weights of the reservoir (13) should take into account additional working fluid inflow G8. We will write down this equation as:

$$r\left(\frac{dm''_{fR}}{d\tau} - G_4x_4 - G_8\right) = k_R(t_1 - t_R) + (\alpha_{ext})_R(t_{ext} - t_R)S_R + (\alpha_{in})_R(t_4 - t_R)S_4 + \cdots$$

$$\dots - C_R\frac{dt_R}{d\tau} + \dots - G_4c'_4(t_R - t_4) \tag{18}$$

The basic problem of calculation is related with determination of the factors ζ_{V2} and ζ_{V8} of hydraulic resistance for equations (16)-(17). Values of these factors could be changed and they are depending on the valve position.

Let's assume that y is the axis along which the valve moves. The y_1 and y_2 are extreme positions corresponding the completely closed and completely open bypass valve. Then range of the change of the valve is $y_{\le}y\le y_2$. We will assume that factor of the valve local resistance is inversely to degree of the valve opening, i.e. $\zeta_8 = 1/A_1(y - y_1)$, where A_1 is a numerical factor. In view of it from the equation (16) we will receive:

$$G_8 = \left(A_1(y - y_1)\left[(p_2 - p_6)2\rho''\omega_8^2\right] + G_8^2\frac{\omega_8^2}{\omega_2^2}\right)^{0.5} \tag{19}$$

Let us assume that gas pressure in bellows is proportional to the absolute temperature, and the value of valve movement depends linearly on pressure difference between the bellows and loop. Then

$$\frac{A_2}{y}(t_9 + 273) - p_S = A_3(y - y_{0)} \tag{20}$$

Where y_0 is the coordinate of the valve position at equality of pressure between the bellows and loop. In case of ideal gas in bellows $A_2 = mR/S$, where m is gas weight in bellows, R is the gas constant, S is the area of cross-section section bellows and lastly A_3 is the factor of bellows stiffness. From last equation it follows that

$$y = \frac{1}{2}\left(\frac{p_S S}{A_3} - y_0\right) + \left(\frac{1}{4}\left(\frac{p_S S}{A_3} - y_0\right)^2 + \frac{A_2 S}{A_3}(t_9 + 273)\right)^{0.5} \quad (21)$$

By solving the equations (18), (19) and (21) together with the equations system (1)-(16) it is possible to determine the valve position, working fluid rate in bypass line and the all LHP characteristics. Operation of the valve - regulator is determined by the following numerical characteristics: A_1, A_2, A_3, y_0, y_1, y_2. At the using inert gas in bellows and absence of heat supply to the bellows these factors are constants. In case of use of two-phase agent and heat supply some of them, and first of all could be changed. It is resulted in changing parameters of regulation.

Figure 9 shows the calculation results of LHP parameters change at the cooling temperature 25^{0}C and at heat generation change in the pump module head. Behaviour of power change is presented in Figure 9. Initial value of the power is 15 W. Gradual decrease of the power by 2 W was observed in every 20 seconds. After reaching 5 W power was increased up to 15 W. In a minute the process was repeated. In accordance with the results pressure regulator provides high quality of saturation temperature stabilization in LHP (curve 4).

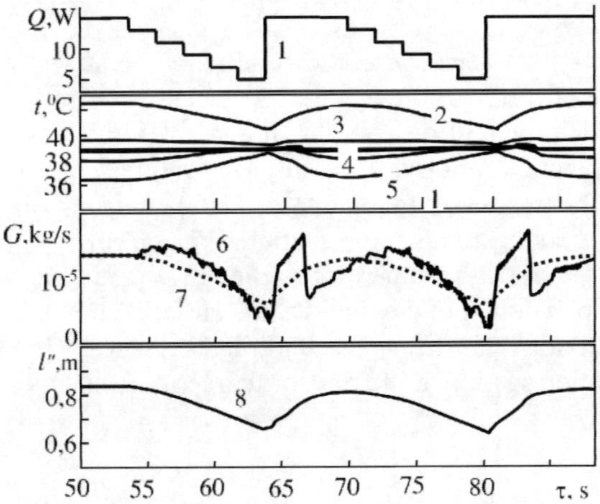

Figure 9. Change in time of LHP parameters when changing the evaporator's heat load (curve 1), 2 - 5 - temperature of the cooled object, evaporator, saturation in the evaporator, saturation in the CP, at the output of the condenser; 6 and 7 - mass flow rates of vapor at the outlet of the evaporator and liquid at the inlet to the evaporator; 8 - length of the condensation section in the condenser.

Stabilization of evaporator body temperature (curve 3) is quite reasonable too. However, the pump module head temperature (curve 2) is changed within two degrees due to the thermal contact and heat capacity and that is above the required normal twice. Working fluid rate at the evaporator outlet (curve 6) and working fluid rate at the compensation chamber inlet (curve 7) are equal only in the set mode and may be significantly different when the power is changed that leads to filling condenser with liquid or working fluid removal from condenser (curve 8).

Changing of the clearance y that characterizes the bypass valve opening is presented by the curve 11 in the bottom diagram. Decreasing the power the valve clearance is decreased too, and at the moment of the power increase the complete closing of the bypass valve during about 30 seconds is occurred.

Thermal-hydraulic auto-oscillations are possible during LHP operation. Fig. 9 shows the example of the steady LHP operation after single cooling temperature decrease (curve 7) from 0 to –5^{0}C. Steady oscillation process with the oscillation period of 100 seconds is observed. It should be noted that there is no total closing of the bypass valve. Oscillations are caused by interaction of thermal and hydrodynamical processes. These oscillations lead to oscillations of the pump module head temperature. Its temperature is within the range of ±0.2^{0}C.

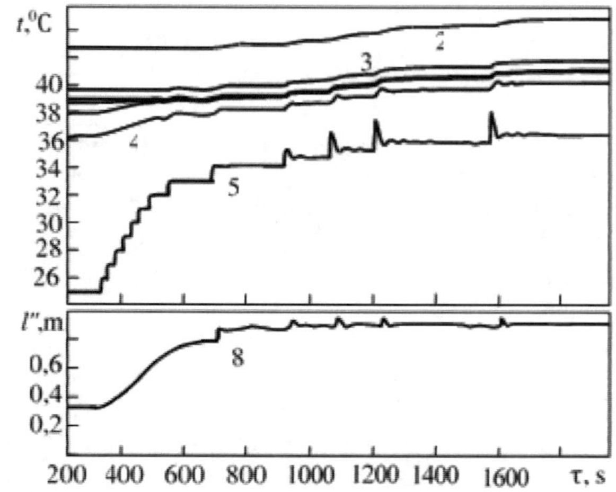

Figure 10. Change in time of LHP parameters with a change in the temperature of the cooling medium (designation of the lines, see Figure 9).

Calculations showed that with a decrease in the cooling temperature by 1 degree every 10 seconds due to the regulator, the temperature of the plate does not practically change, while the cooling temperature decreased by more than 30 degrees. With a stepwise increase in temperature by 1 degree (Figure 10), reverse phenomena are observed which have features. At a cooling temperature of more than 34°C (and adjusting the regulator to maintain 42°C), a qualitative change in the occurring phenomena occurs. At low cooling temperatures, the LHP control was performed by changing the position of the regulator valve. At high temperatures, the valve is normally closed, and self-regulation is provided by varying the supercooling of the liquid at the outlet of the condenser. If at low temperatures a monotonic opening of the condenser occurs (line 8 in Fig. 9), then at a high temperatures rapid change in the length of the condensation section at the time of temperature change.

4. Conclusions

The presented model of the LHP dynamics gives physically grounded calculation results and is consistent with the test results.

The installation of the bypass line and the regulating valve in the LHP allows to significantly improve the quality of thermo-stabilization in comparison with the traditional methods, with no energy costs for regulation.

Presence of two free border of phases (in the condenser and in a compensatory camber) can cause moving of a liquid to a direct and return direction at insignificant additional influence. Calculation results is shown, that under certain conditions there can be auto-oscillations of the liquid, accompanied moving of vapor from the condenser to reservoir and return liquid flow from reservoir.

The cyclic heat input can provoke increase thermo-hydraulic auto-oscillations of the working fluid in a loop that can lead to growth of thermal resistance in 1.5 times and to essential rise in temperature of the evaporator. Thus, the LHP continues to work, but essential growth of temperature in comparison continuous heat load in tests can wrongly be treated, how output LHP out of operation.

During the settlement analysis the opportunity of failure LHP owing to arising auto-oscillations is not established at a cyclic supply of thermal loading. In all cases the LHP keeps working ability though integrated characteristics can essentially different.

5. Nomenclature

A_3	bellows stiffness coefficient (kg/m)
b	perimeter of the heat-exchanger surface (m)
C	total heat capacity (J/K)
c	specific heat capacity (J/(kg·K))
d	diameter (m)
G	mass flow rate (kg/s)
i	specific enthalpy (J/kg)
k	heat transfer coefficient (W/m^{2}K)
m	mass (kg)
p	pressure (Pa)
R	gas constant; radius of pores (m)
r	latent heat (J/kg)
S	area (m^2)
Q	heat flow (W)
q	specific heat flow (W/m^3)
q_l	linear density of heat flow (W/m)
t	temperature (^{0}C)
u	velocity (m/s)
V	volume (m^3)
x	mass vapor content
z	coordinate along the loop (m)
α	heat exchange coefficient (W/(m^{2}K))
λ	heat conductivity (W/(m·K))
ρ	density (kg/m^3)
τ	time (s)
ξ	coefficient of hydraulic resistance
σ	surface tension (N/m)
ω	cross-sectional area (m^2)

Subscripts:

ext	external
f	working fluid
j	sequence number of the section (1 - evaporator, 2 - vapor line, 3 - condenser, 4 - liquid line);
in	internal
w	wall
′	liquid parameters
″	vapor parameters
R	reservoir
s	saturation parameters

6. References

[1] Maydanik Y.F., 2005 Loop heat pipes, *Applied Thermal Engineering* **25** 635-657 .

[2] Goncharov K.A., Kochetkov A.Yu., Buz V.N., 2005 Development of loop heat pipe with pressure regulator, Heat Pipe, Heat Pumps, Refrigerators. *Proceedings of the VI Minsk Int. Seminar.* Belarus 167-172.

[3] Sasin V., Zelenov I., Zuev V., Kotliarov E. 1990 Mathematical model of a capillary loop heat pipe with a condenser-radiator, *20th ICES, Williamsburg*, Virginia.

[4] Bienert W. and Wolf D., 1995 Temperature Control with Loop Heat Pipes: Analytical Model and

Test Results, Proceedings of the *1st International Heat Pipe Conference*, Los Alamos National Laboratory.

[5] T. Kaya and T. T. Hoang, 1999 Mathematical Modeling of Loop Heat Pipes and Experimental Validation, *Journal of Thermophysics and Heat Tranfer* **13**, 3 314-320

[6] Cullimore B. and Baumann J. 2000 Steady State and Transient Loop Heat Pipe Modeling, *SAE Technical Paper Series*

[7] Buz.V., Goncharov K. 2002 Modeling of LHP Performances by Means of Specialized EASY Package Program, *Proceedings of the 12th Int. Heat Pipe Conference, Moscow* 182-187.

[8] Vlassov V.V., Riehl R.R., Modeling of Loop Heat Pipe for Ground and Space Application, *35th International Conference on Environmental Systems, Rome*, Italy, paper 2005-01- 2935.

[9] Chernysheva M.A. and Maydanik Y.F. 2012 Simulation of thermal processes in a flat evaporator of a copper-water loop heat pipe under uniform and concentrated heating, *International Journal of Heat and Mass Transfer* **55** 25-26.

[10] Boubaker R., Platel V. 2016 Dynamic Model of Capillary Pumped Loop with Unsaturated Porous Wick for Terrestrial Application, *Energy* **111** 402-413.

[11] Launay S., Platel V. et al. 2007 Transient Modeling of Loop Heat Pipes for the Oscillating Behavior Study, *Journal of Thermophysics and Heat Transfer* **21**(3), (487-495.

IOP Conf. Series: Materials Science and Engineering 1139 (2021) 012007 doi:10.1088/1757-899X/1139/1/012007

Experimental investigation of a capillary-assisted high performance evaporator

Alexandre Marie[1,2*], Pascal Lavieille[1], Marc Miscevic[1], Laurent Lachassagne[2], Vincent Dupont[2]

[1] LAPLACE, Université de Toulouse, CNRS, INPT, UPS, France
[2] IRT St Exupéry, Toulouse, France

Abstract. A particular architecture for a miniaturized evaporator is proposed in which liquid is spread on the heat source using a thin capillary structure connected to two manifolds. The vapor produced by the heat source is extracted from the device using a separate outlet. A prototype has been designed, manufactured and integrated in a mechanically pumped loop in order to evaluate the evaporator efficiency in terms of heat transfer coefficient and induced pressure drop. In this paper, a capillary structure consisting in parallel rectangular micro-grooves is investigated with HFE7000 used as the working fluid. Results show that the evaporator is able to extract heat flux of the order of 10 W.cm^{-2} with heat transfer coefficients close to 10^4 W.m^{-2}.K^{-1} while inducing pressure losses below a few hundred Pascal. Direct visualizations of the flow in the capillary structure allowed a precise characterization of the evaporator functioning mode, in particular the existence of two distinct two-phase flow regimes depending on the flooding of the capillary structure: nucleate boiling and liquid films evaporation. In order to predict the occurrence of these operating modes, a hydraulic model of the evaporator assembly has been developed and validated thanks to dedicated experimental campaigns. In particular, this model is able to predict the separation between the nucleate boiling and evaporation mode and highlights the fact that the capillary limit is not the main restricting factor in terms of the device maximum manageable heat flux.

1. Introduction

In addition to the well-known current context of miniaturization of heat producing units which leads to rising heat flux to be handled, the increase of the total number of heat sources in a given system appears as an upcoming problematic to consider. In particular, the “more electrical aircraft" target investigated by aircraft manufacturers implies a significantly higher number of dissipating electronic components embedded in a given plane. In practice, this implies the necessity to develop lightweight and compact heat management devices which can easily be embedded in complete cooling systems such as two-phase loops. Then, numerous experimental and theoretical studies have been led on ow boiling in micro-channels evaporators [1,2] This approach, despite being promising in terms of heat transfer coefficient and maximum heat flux density, has two major drawbacks: first, for a given dissipated heat flux, the thermal performances of these devices highly depend on the liquid mass ow rate supplied to the device.

In particular, small outlet vapor quality has to be maintained in order to maximize the evaporator heat transfer coefficient [3,4]. Moreover, their simple architecture induces significant pressure drops as

* corresponding author: marie@laplace.univ-tlse.fr

two-phase ow has to be forced into very small cross-sectional area by the pumping source of the loop. In the case of a secondary evaporator embedded in a capillary pumped loop, this might imply a reduction of the operating range of the complete system. For a mechanically pumped loop, the necessary pumping power may be significantly increased.

To address this issue, a concept of capillary-assisted evaporator has been developed by the ACT company [5]. Their prototype is designed to passively extract liquid from a primary ow using local capillary pumping. The collected ow rate is equal to the vaporization ow rate necessary to extract the applied heat ux. The vaporized ow and liquid overflow are then evacuated through two separated outlets. The use of local capillary pumping allows the stabilization of the liquid/vapor interface as well as self-adaptation of the device. Moreover, the pressure drop induced by the device can be negligible, as the ow through small cross-sections (in the capillary structure) is solely driven by capillary forces while the mechanically pumped ow can be maintained in arbitrary wide sections. Since this proof of concept, this design has been more thoroughly investigated.

On one hand, Park & Crepinsek [6] led an experimental study on a prototype similar to the one proposed by ACT. In their setup the vapor produced by the evaporator is directly sent to the loop reservoir in which it is condensed. By varying the liquid flow rate supplied to the evaporator, the authors showed the existence of an optimal range of this parameter. Indeed, the main liquid ow rate has to be high enough to prevent ow boiling in the liquid supply line but low enough to avoid the evaporator flooding. More recently, Jiang et al. [7] proposed a different loop architecture in which the vapor produced by the evaporator is condensed and then reinjected in the main liquid line. Likewise, a minimum liquid flow rate necessary to ensure proper functioning of the evaporator is highlighted. The start-up phase of the evaporator has also been investigated [8].

Finally, Setyawan et al. [9] proposed a slightly different architecture. Their system, called a hybrid loop heat pipe with pump assistance, is based on the following principle: in usual condition, the loop operates as a regular LHP (in other words, the fluid motion is solely induced by capillary pumping at the evaporator). However if dry-out is detected, a mechanical pump placed upstream of the evaporator is turned on in order to push liquid back into the porous wick. The authors then demonstrate the ability of their system to increase the maximum heat flux that can be handled by the loop. All the hybrid evaporators architectures proposed above are based on the utilization of a \thick" porous media in which, similarly to a standard capillary evaporator, the capillary pumped liquid ow rate and the heat flux applied are in opposite direction. This implies the existence of a vapor layer between the liquid/vapor interface and the heat source and does not promote miniaturization.

In this context, an alternative capillary-assisted evaporator architecture has been developed and tested. In particular, the phase separation between the main liquid ow and the vaporized flow is ensured in a way allowing direct spreading of the liquid on the heated surface. The first objective of this paper is to detail the functioning mechanism of this evaporator as well as its performance in terms of heat transfer coefficient and induced pressure drop. In a second step, a hydraulic model aiming to predict the operating mode of the evaporator in function of two experimental parameters is presented and compared to the experimental data.

2. Experimental setup

2.1 Capillary-assisted evaporator

The evaporator is based on a planar capillary structure. In this study, the considered porous media is a silicon substrate in which 355 rectangular micro-grooves (40 μm wide and 75 μm deep) separated by 16 μm thick walls have been etched using a standard Deep Reactive Ion Etching (DRIE) process. The capillary structure covers a 2 x 2 cm^2 area corresponding to the heated section of the evaporator. Then, as represented on Figures 1 and 2, two liquid manifolds (called "distributors") are placed directly onto the capillary structure. A 2 mm wide groove is etched on the bottom of the distributors to allow the liquid to ow into the capillary structure. The assembly is set into a hermetic container ensuring a pressurized contact between the capillary structure and the distributors as well as the connection of the distributors to the liquid inlet and outlet of the evaporator. The generated vapor is evacuated through a dedicated outlet placed on one side of the evaporator in order to be reinjected in the liquid line downstream of the evaporator. Finally, a sapphire window placed on top of the evaporator allows direct observations of the internal ow distribution. The resulting prototype assembly is represented in Figure 4. Similar to the existing capillary-assisted evaporators mentioned in introduction, this architecture allows a passive regulation of the device, as the capillary structure only collects the fraction of the liquid ow needed on a main liquid line passing through the evaporator.

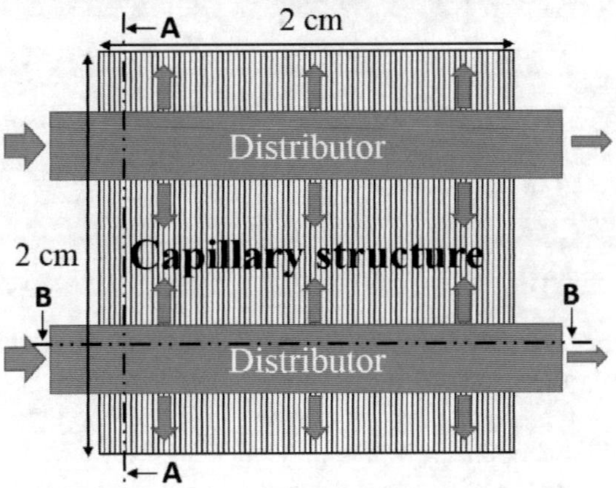

Figure 1 Top schematic view of the distributor/capillary structure assembly.

On the other hand, the major difference of this devices with the existing literature consists in the phase separation between liquid and vapor. Here, instead of a “thick" porous medium ensuring both phase separation and capillary pumping, the flow separation is ensured by the distributors, implying that the unique function of the capillary structure is to extract the needed liquid flow rate and to spread it on the heated surface.

2.2 Test bench

A test bench dedicated to this evaporator, schematically represented in Figure 4, has been designed and set up. It consists of a mechanically pumped two-phase loop in which the ow circulation is ensured by a volumetric pump (gear pump). A copper heating block placed directly in contact with the evaporator, is used as the heat source.

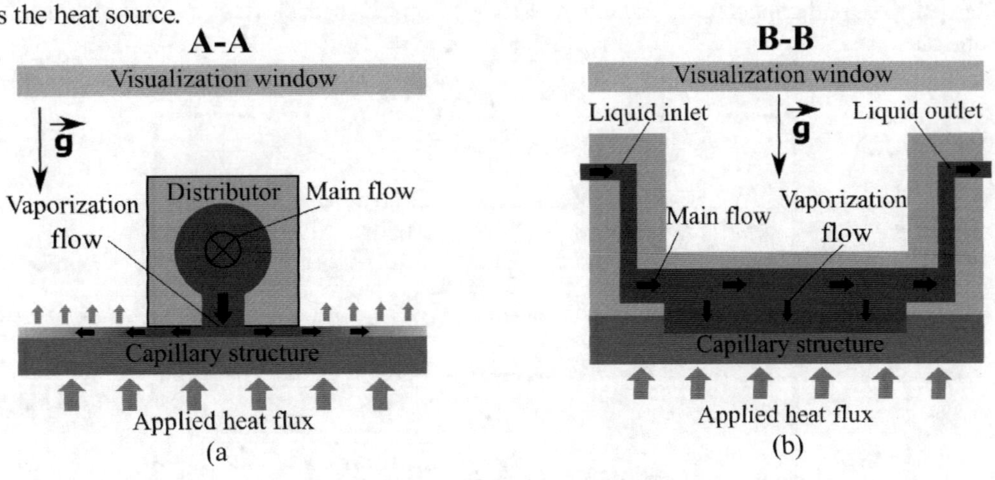

Figure 2 Schematics of the distributor/capillary structure assembly: (a) transversal cross section (left); (b) longitudinal cross section (right).

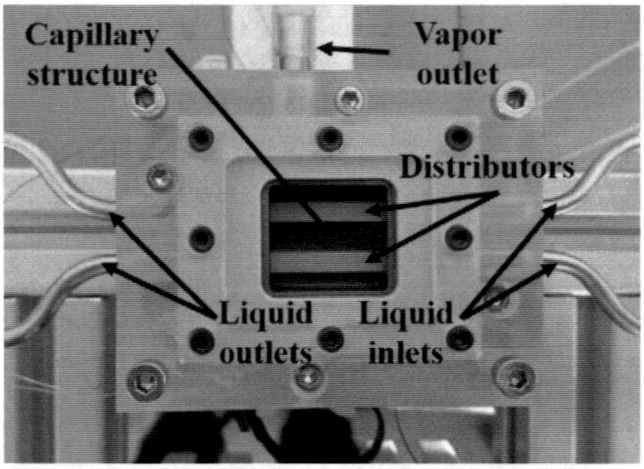

Figure 3 Top view of the evaporator prototype.

A control valve, placed between the evaporator liquid outlet and the liquid and vapor line connection allows the adjustment of the liquid pressure in the evaporator whatever the liquid flow rate of the loop. Indeed, as explained thereafter, this parameter has a strong influence on the functioning of the evaporator. The liquid/vapor mixture is then condensed in a counter-current heat exchanger connected to an external water circuit. After its condensation, the fluid passes through a mass-flowmeter. Finally, a constant pressure reference is set directly downstream of the pump by a two-phase reservoir linked to a thermostatic bath. The working fluid used is degassed HFE7000.

In order to evaluate this two-phase device in terms of cooling capabilities, the following instrumentation is set up. The liquid flow rate imposed by the pump is measured using the mass flowmeter downstream of the condenser (micro-motion CMFS010M, accuracy $\pm 2.10^{-3}$ kg/h, range 0-8 kg/h). The absolute pressure at the entrance of the evaporator is measured by a Keller PA-33X pressure transducer (accuracy ±200 Pa, range 0-2 bar). Two differential pressure transducers Keller PD-33X (accuracy ±25 Pa, range 0-0.2 bar) are used. The first one measures the pressure drop between the liquid inlet (Figure 4, point A) and the liquid/vapor junction (Figure 4, point C) while the other measures the pressure drop between the liquid inlet and the vapor outlet (Figure 4, point B). Calibrated 0.5mm K-type thermocouples (accuracy ±0.2°C) are placed directly in the fluid flow at each inlet and outlet of the evaporator in order to measure the characteristic fluid temperatures. Finally, the evaporator wall temperature is measured with five 0.25 mm calibrated K-type thermocouples (accuracy ±0.2°C) bonded in dedicated grooves machined in the silicon substrate.

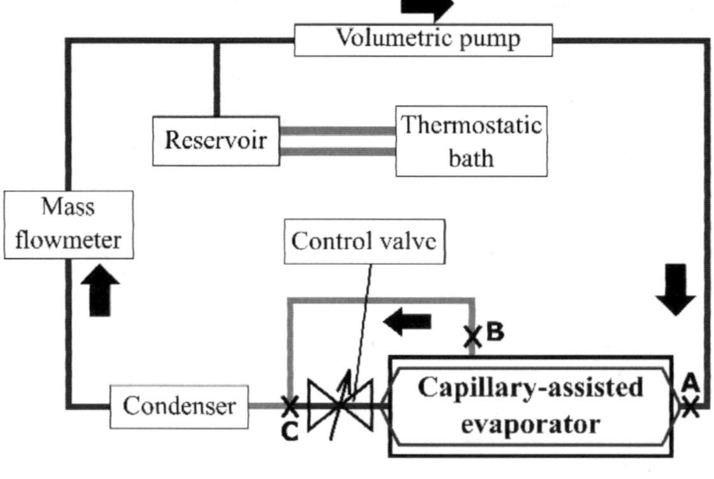

Figure 4 Schematic view of the test bench.

3. Evaporator operation and associated performance

3.1 Qualitative flow distribution

In this section, the control valve setting as well as the liquid flow rate supplied to the evaporator are set constant. Then, the applied heat flux is progressively increased from the onset of phase change in the evaporator until dry-out. During this sequence, the evaporator goes through 4 different operating modes:

- at low heat fluxes, the capillary structure appears overfed with liquid, in other words liquid freely flows from the distributor to the heated surface (Figure 5), thus the heat flux is extracted through nucleate boiling on the capillary structure. At the lowest heat fluxes, a part of the liquid flow directly exits the evaporator through the vapor outlet;
- at higher heat flux, the liquid is constrained within the capillary structure (Fig 6), allowing its capillary pumping from the distributor. As nucleate boiling is no longer observed, heat transfer then occurs through evaporation of the liquid films in the grooves, similarly to the evaporator area of a heat pipe;
- between these two functioning ranges (i.e. at intermediate heat fluxes), a transition mode is observed in which the nucleate boiling and evaporation modes are alternatively observed;
- finally, when a sufficiently high heat flux is applied, the free surface of the capillary structure is completely dried-out (Figure 7). The evaporator wall then achieves a significantly higher equilibrium temperature. As in these conditions the fluid is almost entirely confined under the distributors, direct visualizations of the phase structuration is impossible. Hence the flow repartition described in the right part of Figure 8 (flow boiling in the distributor groove) cannot be observed. However, the observation of vapor in the liquid outlets transparent tubes confirms this hypothesis.

3.2 Evaporator performance as a function of the applied heat flux

A situation similar to the one presented above is considered: the liquid flow rate supplying the evaporator is set at $\dot{m}_{tot} = 3{,}0\ \mathrm{kg.h^{-1}}$ and the control valve is set to induce a pressure drop $\Delta P_{AC} \approx 100$ Pa. The heat flux is then gradually increased from 3.4 to 9.8 W.cm^{-2}. For each measurement point, acquisition of data starts once permanent regime has been reached. Considering the heated surface S of the evaporator, $\overline{T_w}$ the mean of the 5 wall temperatures measurements and Φ_{losses} the evaporator heat loss to the ambient (measured by progressive heating of the vacuumed evaporator), the apparent heat transfer coefficient of the evaporator, is then defined as:

$$h_{app} = \frac{\Phi - \Phi_{losses}}{S \times (\overline{T_w} - T_{sat})} \tag{1}$$

The variation of this parameter with the applied heat flux is represented on Figure 8 with the corresponding functioning modes. Considering the points A, B and C represented on Figure 4 the measured pressure drops ΔP_{AC} and ΔP_{AB} as well as the deduced vapor pressure drop $\Delta P_{BC} = \Delta P_{AC} - \Delta P_{AB}$ are represented on Figure 5. In terms of heat transfer coefficient (

Figure 9), the nucleate boiling mode appears as mildly less effective than the evaporation mode. Indeed, given the manufacturing process of the capillary structure (plasma etching of a plain silicon surface), the heated surface is likely very smooth, inducing a small number of nucleation sites, hence an overall poor nucleate boiling performance. On the other hand, a sharp increase of h_{app} is observed during the transition to evaporation mode. Once stable evaporation is achieved, the heat transfer coefficient remains stable. On the contrary, once dry-out of the free surface occurs (partial at the penultimate point shown on and complete at the last one), the heat transfer coefficient is divided by approximately 5 which seems consistent with the significantly reduced (but not suppressed) wetted area in the assumed flow distribution.

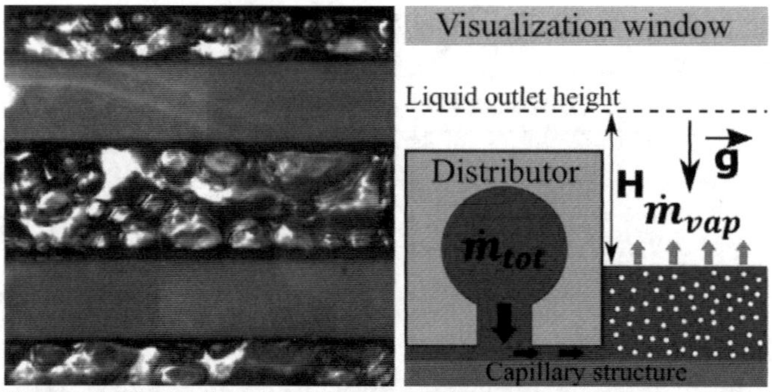

Figure 5 Flow visualization and schematic representation of the phase distribution during nucleate boiling mode.

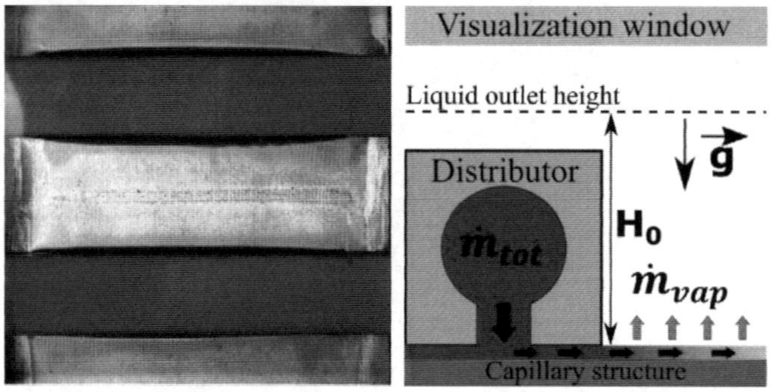

Figure 6 Flow visualization and schematic representation of the phase distribution during liquid films evaporation mode.

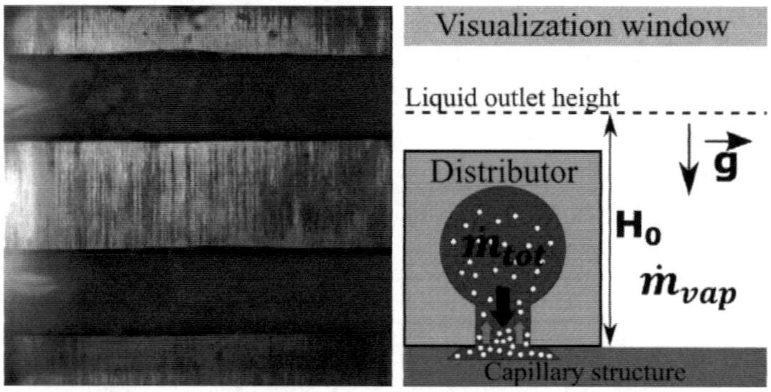

Figure 7 Flow visualization and schematic representation of the assumed phases distribution at dry-out.

Considering pressure drops (Figure 8 b), different tendencies can be observed. On one hand, apart from the dry-out points, the complete liquid line pressure drop (ΔP_{AC}) only slightly decreases with the heat flux. Indeed, as a larger portion of the liquid flow rate is extracted from the main line by the evaporator (to absorb the increased applied heat flux) a lower flow rate (thus, a lower pressure drop) is expected between the liquid outlet and point C. Once dry-out occurs, this pressure drop increases significantly as a result of the transition to two-phase flow in the liquid lines.

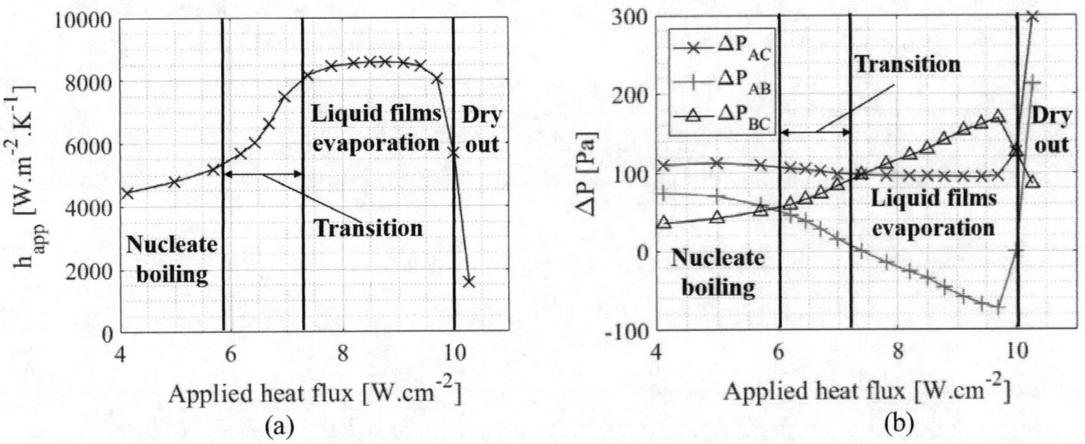

Figure 8 Apparent heat transfer coefficient of the evaporator as a function of the applied heat flux (a), pressure drops induced by the different parts of the evaporator assembly as a function of the applied heat flux (b).

On the other hand, the vapor outlet pressure drop (ΔP_{AB}) progressively decreases from the onset of boiling to the dry-out. Two successive phenomena can explain this variation: At first (in boiling mode), the diminution of the mean liquid level in the evaporator decreases the hydrostatic pressure difference between A and B. Then, during evaporation mode, as the capillary pumped flow rate in the micro-grooves rises, the associated menisci curvatures increases concurrently. As a consequence, a higher capillary pressure is induced leading to a decrease of the measured pressure difference. As capillary pumping is likely to be suppressed once dry-out occurs, the pressure drop then increases sharply.

Finally, the resulting vapor line pressure drop (ΔP_{BC}), simply increases with the increase of the vaporized flow rate passing through, its reduction during dry-out indicates that the vapor flow rate in the vapor line is reduced as a significant part of the produced vapor is driven in the liquid line.

4. Analytical model of transition and dry-out

As seen in the previous section, the two-phase flow distribution is strongly influenced by the experimental parameters. The objective of the model developed thereafter is then to explain and predict the different operating modes observed using a simple hydrodynamic model.

4.1 Model presentation

As discussed earlier, starting from point A (Figure 4), two distinct paths can be taken by the liquid, depending whether it passes through the capillary structure (Figure 4, path ABC) or goes directly through the evaporator (Figure 4, path AC). As those two distinct lines have common starting and ending points, the pressure drops generated by those two parallel flows must be the same (in permanent regime). The pressure drop of the liquid line is set by the control valve setting and the liquid flow rate flowing through the AC path. Therefore, the flow in the line ABC has to adapt itself to this fixed pressure drop. The hypothesis developed here is that the different operating modes observed correspond to the only flow configuration allowing the ABC pressure drop to balance the fixed liquid line pressure drop.

4.2 Pressure drops evaluation

In practice, the pressure drop along the ABC path can be divided into four terms:

- the hydrostatic pressure $\Delta P_g = \rho_l g H$, H being the height difference (represented in Figures 5-7) between the liquid outlet and the liquid level in the evaporator, ρ_l the liquid density and g the gravity acceleration;

- the liquid pressure drop $\Delta P_{grooves}$ induced by the liquid flow in the capillary structure;
- the capillary pressure drop $\Delta P_{cap} = \gamma/R$, induced by the interface radius of curvature R in the capillary structure (considering the significant extension of the liquid films, the longitudinal curvature radius is considered infinite). In the case of nucleate boiling, this term is considered equal to 0 as the capillary structure is flooded;
- the sum of regular and singular pressure drop ΔP_{BC} induced by the vapor flow from the vapor outlet (point B) to the liquid/vapor junction (point C).

Then, the hydraulic equilibrium of the two lines is simply given by:

$$\Delta P_{AC} = \Delta P_{grooves} + \Delta P_{BC} - \rho g H - \Delta P_{cap} \quad (2)$$

The first two terms (i.e. viscous pressure drops) can be estimated as a function of the vaporized mass flow rate Φ_{vap}/h_{lv}. For the liquid flow along the grooves, two different areas represented on Figure 9 have to be considered:

- the closed zone (under the distributor), in which the flow section is a closed rectangle;
- the open zone (downstream of the closed zone), in which the liquid is vaporized along the open-top groove.

Given the very low Reynolds number in the micro-grooves (ranging between 1 and 10 for the experimental data presented in this paper), fully developed and laminar flow is assumed. Therefore, the pressure drop in the micro-grooves can be estimated by the following expressions [11]:

$$\Delta P_{grooves}^{closed} = 2\mathrm{Po}_{closed} \times \mu_l U_l \times \frac{L_{closed}}{D_{h,closed}^2} \quad (3)$$

$$\Delta P_{grooves}^{open} = 2\mathrm{Po}_{open} \times \mu_l U_l \times \frac{0.5\, L_{open}}{D_{h,open}^2} \quad (4)$$

In Eqs.3 and 4, Po is the Poiseuille number of the flow through rectangular channels, estimated by Shah & London correlation. D_h is the hydraulic diameter of the grooves, μ_l and U_l are the liquid viscosity and mean velocity. L_{closed} and L_{open} represents respectively the length of the grooves under the distributors and in the vaporization area. The numerical values of these parameters are summarized in Table 1.

The liquid mean velocity is calculated assuming that the vaporized flow rate is equally distributed in each of the 4 flow directions (2 for each distributor) and in the N microgrooves of cross-sectional area A_l :

$$U_l = \frac{\Phi_{vap}}{4N\rho_l A_l \times h_{lv}} \quad (5)$$

With h_{lv} the fluid latent heat of vaporization. Finally, the 0.5 factor multiplying L_{open} results from the inclusion of a linear diminution of mass flow rate along the groove due to its evaporation [11] (in the assumption of a homogeneous heat flux along the groove).

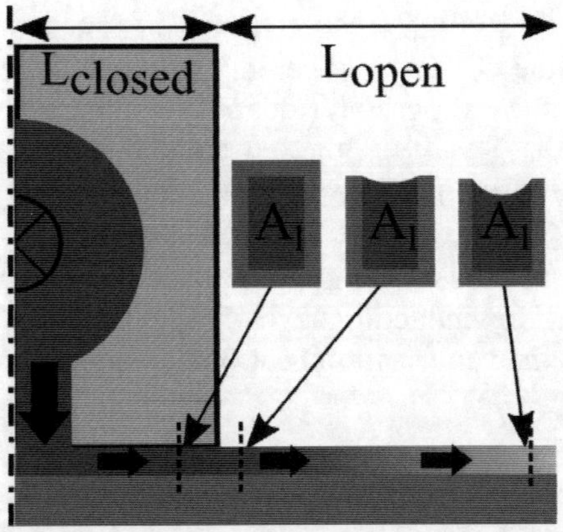

Figure 9 Schematic view of the liquid flow considered along the micro-grooves in evaporation mode.

Table 1 Numerical values of the geometric parameters in Eq. (3), (4) and (5)

	Open zone	Closed zone
Po	18.0	15.3
D_h [µm]	63	52
L [mm]	3	1.5
N	304	304

For the vapor flow, Idel'cik pressure drops memento [10] is used to estimate the pressure drop induced by each hydraulic element of the BC line. For brevity, this calculation is not detailed here, however the complete calculated pressure drop (including the vapor flow in the vapor lines and the liquid flow in the micro-grooves) can be estimated within a 6% error range by the following correlation (for a vaporization heat flux expressed in Watts and a pressure drop in Pascals):

$$\Delta P_{BC} = 0.1802\,\Phi_{vap}^{2}. + 11.62\,\Phi_{vap} - 11.00 \tag{6}$$

As mentioned above, the objective of this model is to predict the flow transitions experimentally observed, in particular the transitions from boiling to evaporation and from evaporation to dry-out. The definition of these two transitions is then the objective of the next section.

4.3 *Transition and dry-out definition*

First, the transition point between boiling and evaporation is defined as the vaporization heat flux for which the grooves are no longer flooded and, in the same time, the meniscus curvature at the entry of the "open zone" is still infinite. In other words, the transition point can be defined as the vaporization heat flux $\Phi_{\mathrm{vap}}^{\mathrm{tr}}$ which satisfies the following equation (derived from Eq. (2)):

$$\Delta P_{AC} + \rho g H_0 = \Delta P_{\mathrm{grooves}}^{\mathrm{closed}}(\Phi_{\mathrm{vap}}^{\mathrm{tr}}) + \Delta P_{BC}(\Phi_{\mathrm{vap}}^{\mathrm{tr}}) \tag{7}$$

Indeed, in the case where $\Phi_{vap} < \Phi_{vap}^{tr}$, the pressure drop generated by the vaporized flow is too small to balance the left hand side of Eq. (7). The only solution is then to reduce the left-hand side by increasing the liquid level in the evaporator (inducing $H < H_0$), in which case nucleate boiling is inevitable. On the contrary, if $\Phi_{vap} > \Phi_{vap}^{tr}$, the left hand side is too small, meaning that a non-zero meniscus curvature at the entry of the open zone is necessary to maintain equilibrium, which implies evaporation along the groove.

Then, the dry-out point is supposed to be imposed by the capillary limit of the evaporator. In other words, it is the vaporization heat flux inducing the maximum meniscus curvature at the end of the "open zone". It can thus be defined as the vaporization heat flux solution of the following equation:

$$\Delta P_{AC} + \rho g H_0 + \Delta P_{cap}^{max} = \Delta P_{grooves}^{closed}(\Phi_{vap}^{max}) + \Delta P_{grooves}^{open}(\Phi_{vap}^{max}) + \Delta P_{BC}(\Phi_{vap}^{max}) \tag{8}$$

4.4 *Experimental validation*

For a given evaporator geometry and capillary structure, the model developed above allows the determination of the vaporized heat flux inducing the boiling/evaporation transition point as well as the maximum vaporization heat flux as a function of the total pressure drop of the liquid line ΔP_{AC}. However, if the value of ΔP_{AC} is directly measured on the test bench, the exact value of the vaporized heat flux is yet unknown as only a part of the heat flux provided by the heating block is absorbed by the vaporization of the fluid. Indeed, an energy balance applied to the evaporator gives the following expression for the heat flux Φ_{tot} dissipated by the heating block:

$$\Phi_{tot} = \Phi_{vap} + \Phi_{AC}^{sensible} + \Phi_{AB}^{sensible} + \Phi_{v}^{sensible} + \Phi_{p} \tag{9}$$

With $\Phi_{AC}^{sensible}$ and $\Phi_{AB}^{sensible}$ the sensible heat fluxes respectively absorbed by the non-vaporized and vaporized flow rates:

$$\Phi_{AC}^{sensible} = (\dot{m}_{tot} - \frac{\Phi_{vap}}{lv}) \times Cp_l(T_l^{out} - T_l^{in}) \tag{10}$$

$$\Phi_{AB}^{sensible} = \frac{\Phi_{vap}}{h_{lv}} \times Cp_l(T_{sat} - T_l^{in}) \tag{11}$$

$\Phi_{v}^{sensible}$ is the heat flux absorbed by the vapor superheating:

$$\Phi_{v}^{sensible} = \frac{\Phi_{vap}}{h_{lv}} \times Cp_v(T_v^{out} - T_{sat}) \tag{12}$$

Φ_p represents the heat loss to the ambient mentioned in section 3.2. In these expressions, Cp_l and Cp_v are respectively the specific heat of the liquid and vapor phases, T_l^{in} and T_l^{out} are the inlet and outlet measured liquid temperatures and T_v^{out} is the measured outlet vapor temperature (at point B). T_{sat} is the saturation temperature of the fluid, deduced from the liquid pressure measured at the evaporator inlet.

Finally, Eq. 9 to 12 have been used to obtain the following expression of the vaporized heat flux:

$$\Phi_{vap} = \frac{\Phi_{tot} - \Phi_p - \dot{m}_{tot} Cp_l(T_l^{out} - T_l^{in})}{1 + \frac{Cp_l}{h_{lv}}(T_{sat} - T_l^{out}) + \frac{Cp_v}{h_{lv}}(T_v^{out} - T_{sat})} \tag{13}$$

Then, an experimental campaign dedicated to this model validation has been carried out: several measurement series similar to the one described in section 3 have been conducted for various values of

IOP Conf. Series: Materials Science and Engineering 1139 (2021) 012007 doi:10.1088/1757-899X/1139/1/012007

ΔP_{AC}. In practice, for each series, the valve setting and liquid flow rate were fixed as a function of the desired pressure drop. In any cases, the liquid flow rate in the loop was set between 1.8 and 3 kg.h-1.

The set of experimental points obtained are then reported on Figure 10 representing Φ_{vap} versus ΔP_{AC}. The two theoretical curves obtained with Eqs. (7) and (8)) are also reported on this figure.

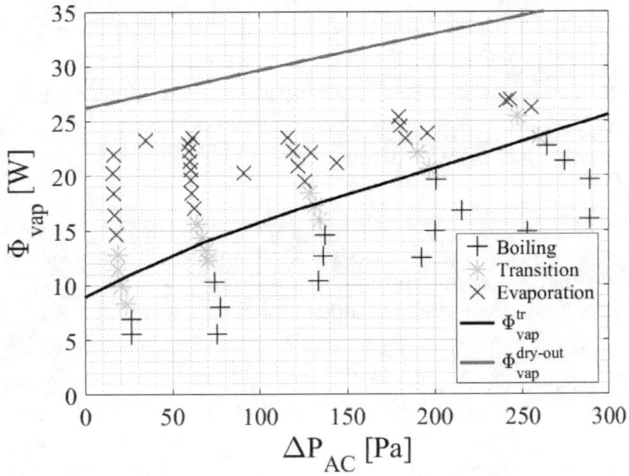

Figure 10 Comparison of the theoretical transition and maximum heat flux (plain lines) with the experimental points (symbols).

On one hand, the model seems to accurately predict the transition between the boiling and evaporation mode. The transition experimental points (star symbols) are found on both sides of the transition curve. As this transition range is not considered by the model (which consider a single transition point), this tendency seems rational. Nevertheless, future work will be necessary to clarify the mechanisms at the origin of this relatively large transition zone (in particular transient phenomena).

On the other hand, the capillary limit predicted by the model seems to be significantly overestimated, in particular for the highest values of ΔP_{AC}. In fact, the maximum heat flux appears weakly dependent of the liquid pressure drop. In practice, this can be explained by the fact that as the evaporation apparent heat transfer coefficient does not vary significantly with the applied heat flux or the liquid pressure drop, the wall superheating linearly increases with the heat flux applied. Hence, the approximatively constant maximum heat flux observed could correspond to the maximum superheating of the wall before the occurrence of nucleation below the distributors discussed previously (Fig.8). As this phenomenon seems to significantly disrupt the capillary pumping of liquid through the micro-grooves, this might lead to a dry-out of the capillary structure before the capillary limit is reached.

5. Conclusions

The design of a prototype and the corresponding experimental data obtained allowed to prove the viability of a particular concept of hybrid evaporator. The qualitative operation of this two-phase device has been highlighted, in particular the existence of two main operating modes imputed to hydraulic equilibrium of the two outlet lines. A specific model has then been developed to predict the two-phase flow repartition in the evaporator as a function of two experimental parameters (pressure drop of the evaporator assembly and vaporized heat flux). This model was able to accurately predict the frontier between evaporation and boiling in function of these two parameters. Nevertheless, several interrogations subsist:

- the physical origin of the transition mode remains to be understood and included in the evaporator functioning model;
- given that the occurrence of boiling during the evaporation mode seems to be at the origin of the maximum heat flux manageable by the evaporator, the resulting dry-out mechanism needs to be clarified in order to accurately predict the dry-out heat flux.

Finally, the transition and dry-out models applied to this evaporator for the same liquid pressure drop range with methanol as a working fluid predict a maximum heat flux two to three times superior to HFE7000 whereas the transition heat flux stays at the same order of magnitude. Thus, in order to investigate the potential of this prototype and to confirm the reliability of the analytical models, further tests with methanol as a working fluid will be performed.

6. References

[1] Tibirica C B, Ribatski G Flow 20113 Boiling in micro-scale channels – synthetized literature review *International Journal of Refrigeration* **36** pp 301-324
[2] Smakulski P 2016 A review of the capabilities of high heat flux removal by porous materials, microchannels and spray cooling techniques *Applied thermal engineering* **104** pp 636-646
[3] Agostini B, Thome J R, Fabbri M, Michel B, Calmi D, Kloter U 2008 High heat flux flow boiling in silicon multi-microchannels – Part I : Heat transfer characteristics of refrigerant R236fa, *International Journal of Heat and Mass Transfer* **51** pp 5400-5414
[4] Agostini B, Thome J R, Fabbri M, Michel B, Calmi D, Kloter U 2008 High heat flux flow boiling in silicon multi-microchannels – Part II : Heat transfer characteristics of refrigerant R245fa, *International Journal of Heat and Mass Transfer* **51** pp 5415-5425
[5] Park C, Vallury A, Perez J 2006 Advanced hybrid loop cooling technology for high performance thermal management, *4th International energy conversion engineering conference San Diego, California, USA AIAA* 4059 pp 4059-4066
[6] Crepinsek M, Park C 2011 Effect of operational conditions on cooling performance of pump-assisted capillary-driven two-phase loop, *Journal of Thermophysics and Heat Transfer* **25** 4 pp 572- 580,
[7] Jiang C, Liu W, Wang HC, Wang DD, Yang YG, Li J, Liu Z C 2014 Experimental investigations of pump-assisted capillary phase change loop *Applied thermal engineering* **71** pp 581-588
[8] Jiang C, Liu W, Liu Z, Yang Y, Bin D, Xiaobing L 2017 Startup characteristics of pump-assisted capillary phase change loop *Applied thermal engineering* **126** pp 1115-1125
[9] Setyawan I, Putra N, Hakim I I 2018 Experimental investigation of the operating characteristics of a loop heat pipe using pump assistance *Applied thermal engineering* **130** pp 10-16
[10] Hopkins R, Faghri A, Khrustalev D 1999 Flat miniature heat pipes with micro capillary grooves, *Journal of Heat Transfer* **121** pp 102-109.
[11] Idel'cik I E 1986 *Memento des pertes de charge* (Ed. Eyrolles, Paris)

IOP Conf. Series: Materials Science and Engineering 1139 (2021) 012008 doi:10.1088/1757-899X/1139/1/012008

A theoretical investigation of a controlled hybrid mechanical/capillary pumped loop

Marie Levêque[12*], Sébastien Dutour[1], Marc Miscevic[1], Pascal Lavieille[1], Yves Bertin[3], Raphaël Mari[24]

[1]University of Toulouse, UPS, LAPLACE, 31062, Toulouse cedex 9, France
[2]IRT-saint Exupéry, 3 rue Tarfaya - CS 34436, 31405 Toulouse cedex 4, France
[3]Institut PPRIME (UPR CNRS 3346), ISAE-ENSMA, 1 av. Clément Ader-BP40109, 86961 Futuroscope-Chasseneuil, France
[4]Airbus Defence and space, 31 rue des Cosmonautes, 31400 Toulouse, France

Abstract A series hybrid mechanical/CPL system is a way to improve the performance characteristics over a plain CPL system, as demonstrated by Schweickart et al [1]. However, few hybrid loops actually operate because the control of the system remains difficult. An investigation of a controlled hybrid CPL is proposed based on a dynamic model, consisting of the coupling of a CPL model and a mechanical pump whose speed is controlled by a PID. We first theoretically determined the controller parameters versus the CPL characteristics in order to optimize the command for a given CPL. In a second part, some simulations of two different architectures were performed and analyzed. These results have confirmed that the hybrid system is very attractive to greatly improve the CPL performance.

1. Introduction

Two-phase capillary pumped loops such as CPL and LHP are passive and highly efficient systems for the cooling of a large range of electronic components. The maximum heat load operation is given by the maximum pressure drop across the wick (the capillary limit) or when vaporization deeply develops in the capillary structure (thermal limit). Since the thermal limit occurs at very high heat load, failures in operating conditions are largely due to the capillary limit. This becomes more and more critical for terrestrial transportation applications in which two-phase loops experience acceleration and dynamic changes of position in the gravity field and in which demanding heat load occur. As explain by Schweickart et al [1], a way of overcoming this limitation is to introduce a mechanical pump which would compensate the pressure drop of the loop. The resulting hybrid mechanical/capillary pumped loop (HCPL) gives thus a lot of new opportunities. For instance, the pump may be used to control the evaporator pressure drop so that gravity is compensated by the mechanical pump and does not affect the capillary pumping. In space industry, this could benefit to thermal characterization tests performed on ground in unfavorable gravity configuration, the maximum heat load being more representative of the one in 0g environment. The same advantages could be provided for the acceleration in terrestrial transportation applications. Moreover, the hybrid system allows to examine the use of some new working fluids with lower Global Warming Potential and toxicity even if the surface tension is low,

* corresponding author: marie.leveque@laplace.univ-tlse.fr

which means a lower maximum capillary pressure. It can also permit to dimension the capillary evaporator for the loop nominal functioning point, and not for rare phenomenon.

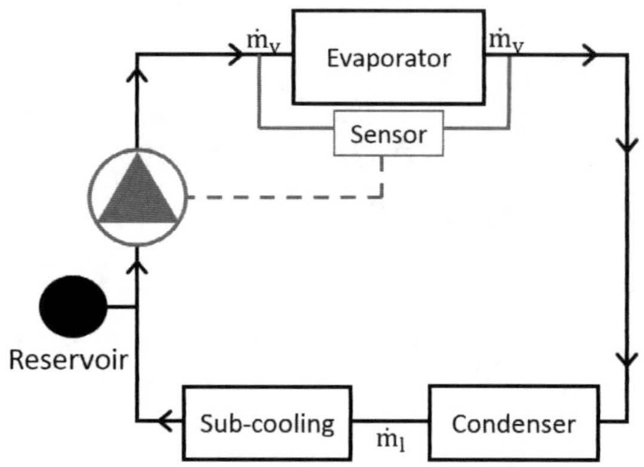

Figure 1. Schweickart et al [1] hybrid loop schematic representation.

If the concept appears extremely fruitful, only a few studies in the literature are related to hybrid systems. The first one was done by Schweickart et al [1] at the NASA. They examined the coupling of a multi-evaporator CPL with a volumetric pump in series located at the inlet of the evaporators. In such configuration (Figure 1), the pump speed has to be controlled as the flow rate is still related to the heat load at the evaporator. They demonstrated the hybrid system operation based on a controller using a differential pressure sensor at the evaporator. A significant but still limited improvement of the performance (14 %) over a plain CPL was obtained. According their results, the fixed liquid/vapor interface disappear with this configuration, the loop becomes a mechanical two-phase loop with an improved evaporator.

More recently, Park et al. [2] and Crepinsek et al. [3] have built a hybrid two-phase loop with a volumetric pump for terrestrial military purpose and heat flux up to 100 W/cm² (Fig. 2). Their hybrid system consisted of a mechanical volumetric pump placed between the reservoir and the evaporator in a parallel configuration. The mechanical pump imposed a high and constant flow rate at the inlet of the evaporator in the first branch of the loop. The capillary wick of the evaporator collects a liquid flow rate accordingly to the heat flux imposed at the evaporator. The vapor then flows into the condenser placed on the second branch. The remaining liquid returns to the reservoir. The main advantage of this system is that it does not required pump control.

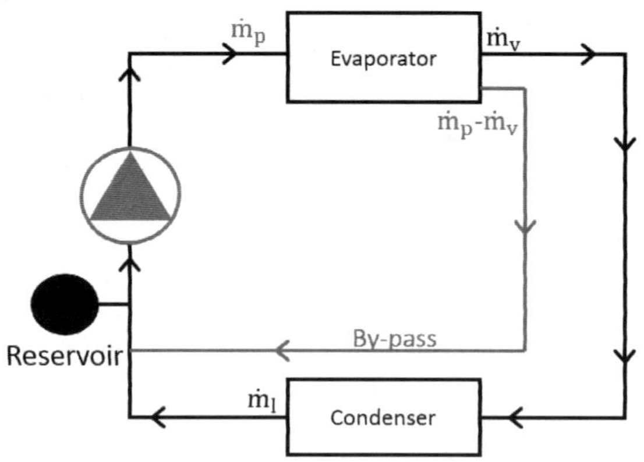

Figure. 2. Park et al [2] hybrid loop schematic representation.

The objective of the Park et al hybrid loop was to increase the maximum heat load by establishing a continuous liquid flow under the wick. This system was also studied by Sarraf and Anderson [4] who demonstrated the system reliability on a LHP for high heat flux and high temperature (up to 200 °C) for electronic application. They proved the system operation up to a heat flux density of 67 W/cm².

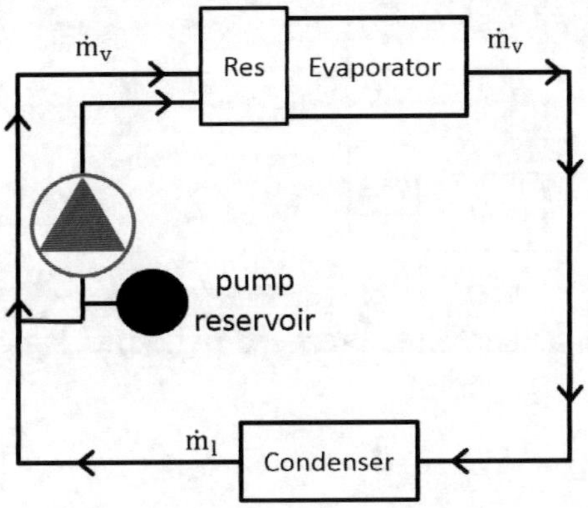

Figure 3. Setyawan et al [5] hybrid loop schematic representation.

Setyawan et al [5] investigated a LHP assisted with a mechanical pump (Figure 3). The aim of their work was to avoid dry-out during the starting of the loop. When the pressure at the evaporator, measured by a sensor, reached a set value, the volumetric pump started and flooded the evaporator. When the mechanical pump was switched off the nucleate boiling allowed the fixation of the liquid/vapor interface. This system has prevented the wick dry-out during starting and transient regime of the capillary loop. In all those studies, the used mechanical pump is volumetric. The coupling of a capillary evaporator with a non-volumetric pump was not yet investigated. We proposed in this work a theoretical and numerical investigation of the behavior and the control of a series HCPL close to the one presented by Schweickart et al [1], but with a non-volumetric pump. The objective is to extend the working range of the loop with a simple architecture while keeping the thermal performance of a capillary evaporator due to the fixed liquid vapor interface. First, the concept is briefly presented in the following section, then, a mathematical modeling is proposed. Numerical results are finally analyzed.

2. Concept

Unlike volumetric pump, a non-volumetric pump such as centrifugal pump is not a flow rate generator. So the flowrate of the system is still fully imposed by the capillary evaporator. The non-volumetric pump could reduce the pressure drop of the loop by generating an opposite pressure difference.

In a first part, two configurations were examined (Figure 4): (i) the mechanical pump is located between the condenser and the reservoir (architecture 1) or (ii) at the inlet of the evaporator down-stream the reservoir (architecture 2). The approach is based on a modelling of the hybrid system consisting in coupling the dynamic model of CPL with the model of a controlled centrifugal pump. Some theoretical efforts were made to investigate the feedback control leading to relations between the controller parameters (PID) and the CPL characteristics (evaporator conductance, heat transfer coefficient at the condenser, pressure drop in line, etc.).

In a second part, some simulations of the controlled hybrid system, including a CPL model that was previously investigated and validated by Kaled *et al* [6], were performed.

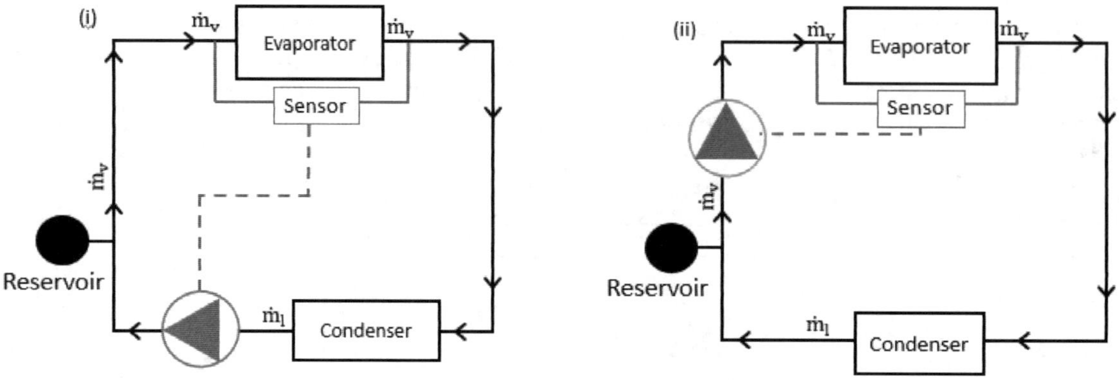

Figure 4. Schematic representation of the proposed hybrid architectures loop: (i) architecture 1 (ii) architecture 2.

3. Modelling

3.1 Model of the HCPL with a pump located at the inlet of the reservoir (architecture 1)
The HCPL model is based on the Dutour and Kaled [7] CPL modelling work. The loop dynamic is mainly controlled by the thermal inertia at the evaporator, the two-phase dynamic at the condenser, the hydraulic inertia in the liquid line and the reservoir dynamic. Considering the reservoir saturation temperature is very well controlled, the loop dynamic can be modelled with only three state equations (1) (2) and (3.1). Main parameters are shown on Figure 5 for architecture 1.

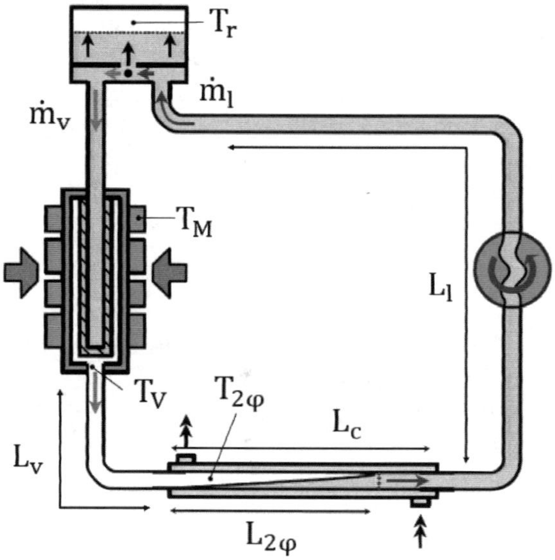

Figure 5. Schematic representation of the main parameters for architecture 1.

The wick and evaporator walls are represented with an equivalent capacitance C. G is the total conductance between the global mass temperature T_M and the vapor temperature, close to the vaporization interface, T_v.

$$\frac{dT_M}{dt} = \frac{Q(t)}{C} - \frac{G}{C}(T_M - T_v) \tag{1}$$

In the condenser, the temperature in the two-phase zone is supposed uniform. The main hypothesis consists in assuming a constant liquid-vapor phase distribution in the condenser during transient phases. It leads to equation (2) with a constant void fraction α.

$$\frac{dL_{2\varphi}}{dt} = \frac{\dot{m}_v - \dot{m}_l}{\alpha A_C(\rho_v - \rho_l)} \tag{2}$$

The liquid inertia is represented in equation (3.1) with a momentum balance between the condenser outlet and the reservoir inlet. $P_{R,low}$ is the pressure at the reservoir inlet and Z_R the height difference between the reservoir and the condenser:

$$\frac{d\dot{m}_l}{dt} = \frac{p_{sat}(T_{2\varphi}) - p_{R,low} - \Delta p_l - \rho_l g Z_R}{\frac{L_C - L_{2\varphi}}{S_C} + \frac{L_l}{S_l}} + \frac{\Delta p_{pump}}{\frac{L_C - L_{2\varphi}}{S_C} + \frac{L_l}{S_l}} \tag{3.1}$$

Some additional assumptions have been done: the flow inertia in the vapor line and in the reservoir are neglected compared to pressure drop. Heat is considered to be consumed in vaporization process, thus specific heat and heat losses are neglected. According to these assumptions, closure equations (4.1) to (4.4) and equation (4.6) are given by Dutour and Kaled [7]. The two-phase condensation coefficient (equation (4.5)) is given by the Wedekins *et al* [8] correlation for low liquid flow rate in an annular condenser. In this correlation a mean void fraction α is considered.

$$\dot{m}_v = G\frac{(T_M - T_v)}{\Delta H_{lv}} \tag{4.1}$$

$$T_v = T_{2\varphi} + \frac{\Delta p_v + \Delta p_{2\varphi}}{a_{vap}} \tag{4.2}$$

$$h_{2\varphi}\pi D_{ce} L_{2\varphi}(T_{2\varphi} - T_{sink}) = \Delta H_{lv}\left[\dot{m}_v - \rho_v\frac{\dot{m}_l - \dot{m}_v}{\rho_l - \rho_v}\right] \tag{4.3}$$

$$\frac{1}{h_{2\varphi}D_{ce}} = \frac{1}{h_{2\varphi,i}D_{ci}} + \frac{1}{h_{1s}D_{ce}} + \frac{ln\left(\frac{D_{ce}}{D_{ci}}\right)}{2\lambda_v} \tag{4.4}$$

$$h_{2\varphi,i} = 0.725^{\frac{4}{3}}\alpha\left[\frac{g(\rho_l - \rho_v)\lambda^3\pi L_{2\varphi}}{\mu\left(\dot{m}_v - \dot{m}_l\frac{\rho_v}{\rho_l}\right)}\right] \tag{4.5}$$

$$p_{R,low} \approx p_{sat}(T_R) \tag{4.6}$$

The pressure drops are considered to vary as the second order of the mass flow rate (equations (5.1) to (5.3)):

$$\Delta p_l = f_{l,1}\dot{m}_l + f_{l,2}\dot{m}_l^2 \tag{5.1}$$

$$\Delta p_V = \Delta p_{groove} + \Delta p_v = f_{v,1}\dot{m}_v + f_{v,2}\dot{m}_v^2 \tag{5.2}$$

$$\Delta p_{2\varphi} = f_{1,2\varphi}\frac{L_{2\varphi}}{L_C}\dot{m}_v + f_{2,2\varphi}\frac{L_{2\varphi}}{L_C}\dot{m}_v^2 \tag{5.3}$$

where f are the friction factors.

IOP Conf. Series: Materials Science and Engineering 1139 (2021) 012008 doi:10.1088/1757-899X/1139/1/012008

3.2 Model of the HCPL with a pump located at the outlet of the reservoir (architecture 2)

For architecture 2, equation (3.1) becomes equation (6). As the pump is no more situated on the liquid line.

$$\frac{d\dot{m}_l}{dt} = \frac{p_{sat}(T_{2\varphi}) - p_{R,low} - \Delta p_l - \rho_l g Z_R}{\frac{L_c - L_{2\varphi}}{S_c} + \frac{L_l}{S_l}} \quad (6)$$

The other state equations and closure equations remain unchanged.

3.3 Mechanical pump and control modelling

The pressure generated by a centrifugal pump depends on the mass flow rate and of its rotational speed $\omega_p(t)$ (equation (7)). To model a perfect mechanical pump, nominal values (which are intrinsic values of the chosen pump) are required. In a second step, equation (6) can be replaced by the pump characteristic given by the pump constructor.

$$\Delta p_{pump} \approx \Delta p_{nom} \left(\frac{\omega_p(t)}{\omega_{p,nom}}\right)^2 - \frac{\Delta p_{nom}}{\dot{m}_{nom}} \left(\frac{\omega_p(t)}{\omega_{p,nom}}\right) \dot{m}_{pump} \quad (7)$$

Then, the time delay between the command set value and the pump set value is defined in equation (8):

$$\frac{d\omega_p}{dt} = \frac{\omega_{p,set} - \omega_p}{\tau} \quad (8)$$

3.3.1 PID controller

To control the HCPL, the output which will be observed, must be defined. The chosen value is the pressure drop at the evaporator liquid-vapor interface Δp_{cap}. The difference between Δp_{cap} and $\Delta p_{cap,\,max}$ can then be determined. Moreover, a feedback loop can be achieved using Δp_{cap} in order to control the pump and keep the evaporator away from the capillary limit.

For architecture 1, the pump is between the condenser and the reservoir. Equation (9) is used to determine Δp_{cap} including the pressure drop in the pump and the inertia effect in the liquid line:

$$\Delta p_{cap} = \Delta p_{grooves} + \Delta p_v + p_{sat}(T_{2\varphi}) - p_{sat}(T_R) - \rho_l g(Z_R - Z_{ev}) + \Delta p_{l,2} + \Delta p_{wick} \quad (9)$$

With $\Delta p_{l,2}$ the pressure drop between the reservoir and the evaporator inlet, Z_{ev} the height difference between the evaporator and the condenser and Δp_{wick} the pressure drop in the capillary evaporator wick. For architecture 2, the mechanical pump is situated between the reservoir and the evaporator. Equation (10) is used to determine Δp_{cap}:

$$\Delta p_{cap} = \Delta p_{grooves} + \Delta p_v + P_{sat}(T_{2\varphi}) - P_{sat}(T_R) - \rho_l g(Z_R - Z_{ev}) + \Delta p_{l,2} - \Delta P_{pump} + \Delta p_{wick} \quad (10)$$

A command law corresponding to a PID controller is then chosen to change the rotational speed of the mechanical pump (equation (11) and (12)) so that Δp_{cap} tends to its set value $\Delta p_{cap,set}$:

$$\omega_{p,set}(t) = K_P.\varepsilon(t) + K_I \int_0^{t_f} \varepsilon(t).dt + K_D \frac{d\varepsilon(t)}{dt} \quad (11)$$

$$\varepsilon(t) = \Delta P_{cap} - \Delta P_{cap,set} \quad (12)$$

The evaporator pressure difference is then controlled to not exceed $\Delta p_{cap,set}$. When the pressure drop in the circuit is lower than this set value, the HCPL works like a CPL, i.e. in a pure capillary pumping mode. Then, when the capillary pressure difference reaches the set value, the mechanical pump is started. The rotational speed of the centrifugal pump is controlled with the PID to keep the evaporator pressure difference constant.

3.3.2 Analytical expressions of the controller coefficients
Dutour *et al* [7] showed that an accurate solution of a nonlinear dynamic model of CPL is provided by a second-order linear system where the undamped natural frequency ω_0 and system damping ratio ξ can be determined as functions of CPL parameters (evaporator equivalent conductance G and equivalent capacitance C, global heat transfer coefficient at the condenser, friction coefficients and fluid properties). Following this approach, the transfer function of a CPL can be defined as:

$$F(p) = \frac{K.\omega_0^2}{p^2+2.\xi.\omega_0.p+\omega_0^2} \tag{13}$$

The second order transfer function for a PID controller is simultaneously defined by:

$$F(p) = \frac{K_D.p^2+K_P.p+K_I}{p} \tag{14}$$

The optimal control coefficients are then determined with parameters ω_0 and ξ and by imposing the proportional coefficient K_p:

$$\begin{cases} K_p = 2.K_D.\xi.\omega_0 \\ K_I = \frac{K_D}{4}\left[\left(\frac{K_P}{K_D}\right)^2 - (2.\xi.\omega_0)^2 + 4.\omega_0^2\right] \end{cases} \tag{15}$$

4. Simulations
Some simulations were performed applying heat load steps at the evaporator.

*4.1 Performance simulations with the mechanical pump at the inlet of the reservoir (*architecture 1)
For architecture 1 (Figure 4.i), the flow rate in the pump is the flow rate in the liquid line:

$$\dot{m}_{pump} = \dot{m}_l \tag{16}$$

Figure 6 shows the mass flow rate in the pump when heat load steps are applied to the evaporator. The liquid redistribution between the condenser and the reservoir leads to large overshoots of the mass flow rate. Thus, in this architecture, the control of the pump is highly influenced by the liquid flow dynamics and the applied power cycle at the evaporator. However as shown in Figure 7, it was possible for this power cycle to find a set of controller parameters leading to a robust control of the HCPL. As explain in paragraph 3.3.1, when the pressure drop in the circuit is lower than this set value, the HCPL works like a CPL, i.e. in a pure capillary pumping mode. Then, when the capillary pressure difference reaches the set value, the mechanical pump is started. The pump rotational speed depends of the heat load maintaining the evaporator pressure difference constant at the set difference pressure ($\Delta p_{cap,set}$ = 9 kPa). As a consequence, the HCPL was able to extend the CPL functioning range up to 2500 W for this simulation i.e. beyond the capillary limit of the wick (1200 W) and up to 40 kPa for the total pressure drop in the circuit. The HCPL maximum heat flux is defined by the maximum pressure difference the mechanical pump can generate.

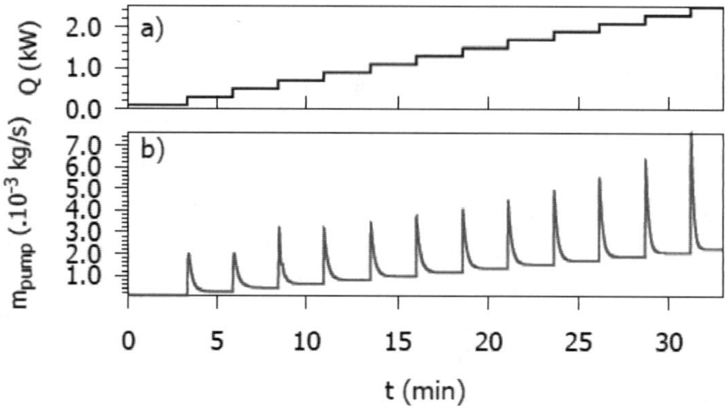

Figure 6. Small amplitude heat flux step at the evaporator from 100 W to 2500 W with architecture 1. a) Power cycle b) flow rate across the pump.

A limitation was observable for this CPL configuration (unfavorable gravity position) when the heat load was decreasing. Actually, during this transient, the liquid flow rate is undershooting and a reversed flow rate could exist in the liquid line. As a consequence, the pressure drop in the pump drastically decreased and the control of the HCPL failed. Some additional considerations on this architecture including for instance a check valve have to be tested.

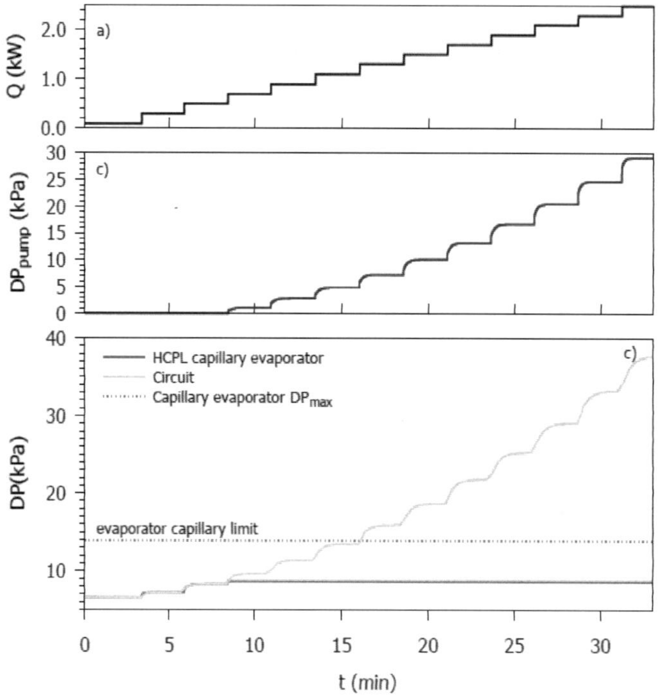

Figure 7. Small amplitude heat flux step at the evaporator from 100 W to 2500 W with architecture 1. a) Power cycle b) Pump differential pressure c) evaporator and circuit HCPL differential pressure.

4.2 Performance simulations with the mechanical pump at the outlet of the reservoir (architecture 2)

In this configuration (Figure 4.ii) the liquid flow rate across the pump is the flow rate due to the vaporization (equation (17)). Actually, as the mechanical pump is a non-volumetric one, the flow rate in this branch is still imposed by the capillary evaporator since the evaporator pressure drop remains positive.

$$\dot{m}_{pump} = \dot{m}_v = \frac{\dot{Q}_{ev}}{\Delta H_{lv}} \tag{17}$$

Figure 8 shows the mass flow rate in the pump when heat load steps are applied to the evaporator. An important benefit compared to architecture 1 is that: (i) There is no overshoot consecutive to an increase of the heat load steps and (ii) the flow rate across the pump is not reversed when the heat load is decreased. This avoids the failure of the HCPL control as described in the previous subsection.

Figure 9 shows the simulation of the HCPL when the evaporator pressure difference is controlled to not exceed Δpcap,set = 11 kPa. As explain in paragraph 3.3.1, when the pressure drop in the circuit is lower than this set value, the HCPL works like a CPL, i.e. in a pure capillary pumping mode. Then, when the capillary pressure difference reaches the set value, the mechanical pump is started. The rotational speed of the centrifugal pump is controlled with the PID to keep the evaporator pressure difference constant. On Fig. 9.c) the well-controlled capillary pressure difference can be observed for both increasing and decreasing heat load steps during the hybrid pumping mode. During this mode, the total pressure drop in the circuit exceeded for 3 times the capillary limit. Returning to low heat loads, the pump is stopped when the pressure difference generated by the centrifugal pump become lower than a set value (here 100 Pa), the loop then works as CPL again.

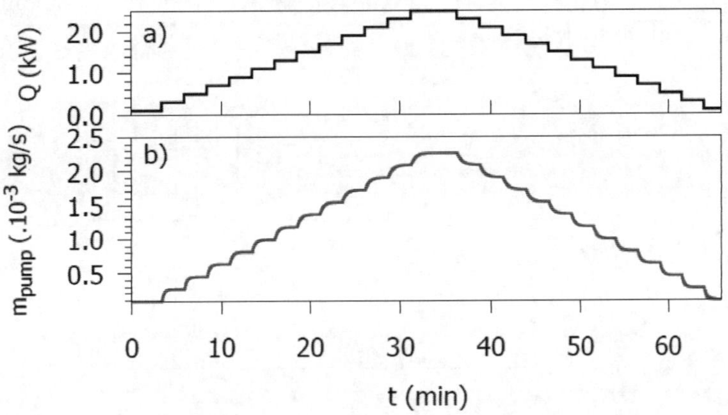

Figure 8. Small amplitude heat flux step at the evaporator from 100 W to 2500 W with architecture 2. a) Power cycle b) flow rate across the pump.

The simulation shows an important enhancement of the pumping performance: the HCPL extended the operating range from 1 200 W to 2 500 W with the chosen mechanical pump and a suitable condenser. As for architecture 1, the HCPL maximum heat flux is defined with the maximum pressure the mechanical pump can generate. This enhancement is also expected in the HCPL response to large amplitude power cycle. We simulated two different HCPL configurations: (i) when the pressure drop in the circuit is dominated by the pressure drop in the vapor line and (ii) when the pressure drops in both lines are comparable. As a consequence, the dynamics in the liquid line is negligible for the first configuration while it is not in the second one.

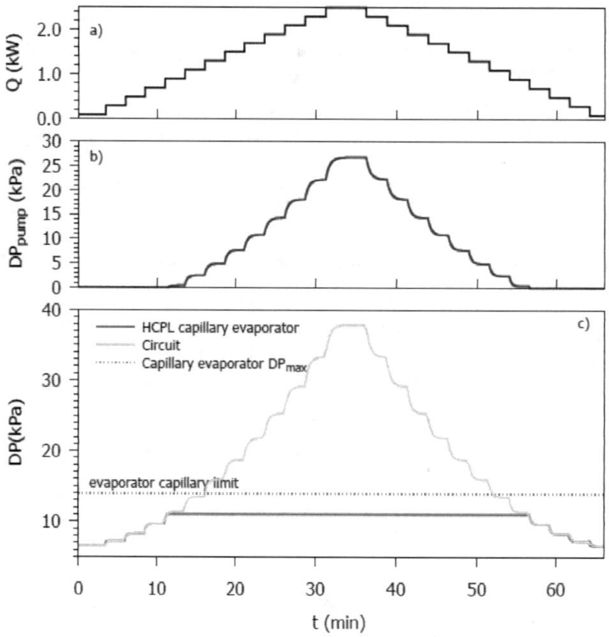

Figure 9. Small amplitude heat flux step at the evaporator from 100 W to 2500 W with architecture 2. a) power cycle b) Pump differential pressure c) evaporator and circuit HCPL differential pressure.

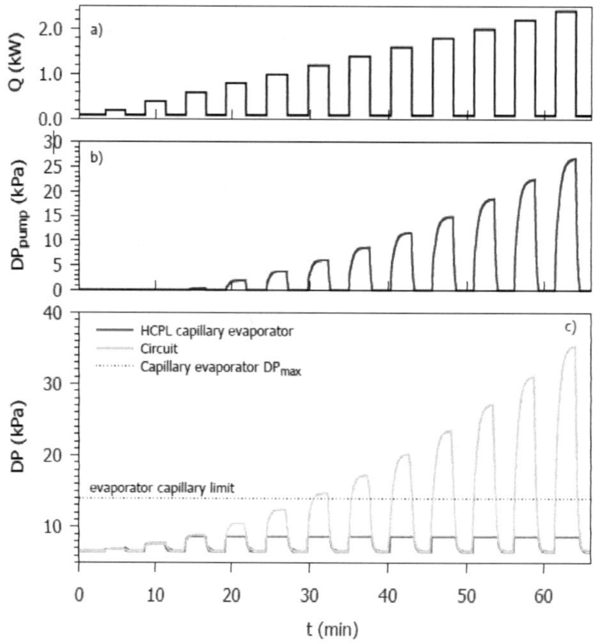

Figure 10. Architecture 2 with strong pressure drop in the vapor line. a) Power cycle b) Pump differential pressure c) evaporator and circuit HCPL differential pressure.

Figure 10 shows the simulation of the dynamics of the HCPL when large amplitude of steps (up to 2 300 W) are applied considering the liquid line dynamics is negligible. The pump starts and stops efficiently for each step and the HCPL works well.

On Figure 11, the simulation was performed with comparable pressure drop in both lines. The consequence is that the pressure drop in the loop is affected by the liquid flow rate overshoot and could exceed the capillary limit during the transient as shown on Figure 11.c. However, the simulations demonstrated that the HCPL control is still robust with a pressure drop set at Δpcap,set = 9 kPa.

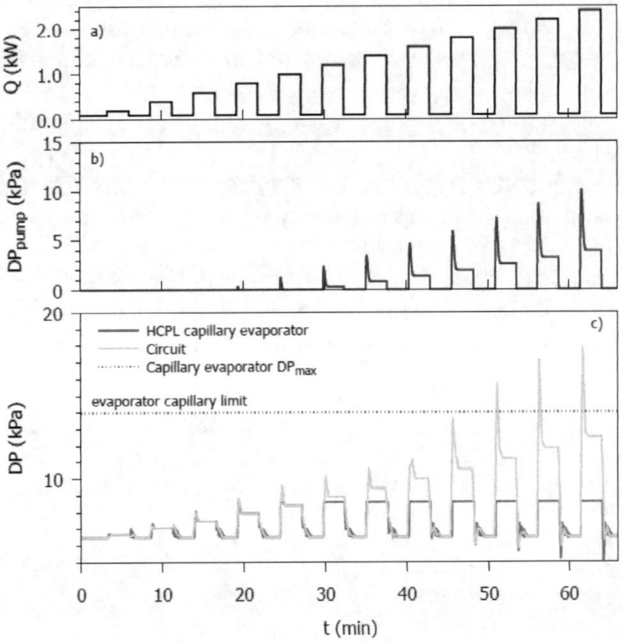

Figure 11. Architecture 2 when HCPL is affected by the liquid line dynamics. a) Power cycle b) Pump differential pressure c) evaporator and circuit HCPL differential pressure.

5. Conclusions

The interest of a hybrid two-phase capillary loop (HCPL) with a non-volumetric pump was investigated in this work. The approach consisted in modelling the coupling between a CPL and a centrifugal pump. The control of the pump (the actuator) is based on PID controller based on a pressure drop sensor. Two architectures were simulated: (i) in architecture 1, the mechanical pump is placed at the inlet of the reservoir, (ii) in the second architecture, the mechanical pump is placed at the outlet of the reservoir. An expression of the PID parameters based on the CPL properties has been established. So that, based on the model, it is possible to find the suitable pump and command for a given CPL.

Analyses of the dynamics of the resulting system showed that this coupling could greatly extend the pumping performance of a CPL. For architecture 1, simulations also showed an improvement for increasing heat load steps. However, when the heat load is decreased, the control failed if a reverse flow occurred in the liquid line. This point has to be addressed and could probably be fixed using a suitable check valve. The best robustness was reached by architecture 2: the improvement of the functioning range of the loop and the control performance is demonstrated for increasing and decreasing heat load, for both large heat load steps and transient step with high flow rate peaks.

6. References

[1] Schweickart R, Ottenstein L, Cullimore B, Egan C, Wolf D 1989 Testing of a controller for a hybrid capillary pumped loop thermal control system. *IEEE,* **69**

[2] Park C, Vallury A, Zuo J, Perez J, Rogers P 2011 Spacecraft thermal management using advanced hybrid two-phase loop technology *AIP Conference Proceedings* **880** pp 11-18

[3] Crepinsek M, Park C 2011 Effect of operational condition on cooling performance of pumped addicted and capillary driven two-phase loop *American institute of Aeronautics* **811** pp 11-18

[4] Sarraf D B, Anderson W G 2015 High Temperature and High Heat Flux Thermal Management for Electronics, *https://www.1-act.com*

[5] Setyawan I, Putra N, Hakim I I 2018 Experimental investigation of the operating characteristics of a loop heat pipe using pump assistance *Applied thermal engineering* **130** pp 10-16

[6] Kaled A, Dutour S, Platel V, Lluc J 2015 Experimental study of a Capillary Pumped Loop for cooling power electronics: Response to high amplitude heat load steps *Applied Thermal Engineering* **89** pp 169–179

[7] Dutour S. and Kaled A 2017 Analytical and numerical results for the dynamics of capillary pumped loops and loop heat pipes subjected to high amplitude heat load steps, *Applied Thermal Engineering* **126** pp 1107-1117

[8] Wedekind G L, Bhatt B L, Beck B T 1978 A system mean void fraction model for predicting varius transient phenomena associated with two-phase evaporating and condensating flows, *International Journal of Multiphase Flow* **4** pp 97-114

Water desalination and power cogeneration utilizing heat pipe heat exchanger

J Velardo[1*], J Searle[1], J Nihill[1], A Date[1] and A Akbarzadeh[1]

[1] School of Engineering, RMIT University, Bundoora, Victoria, 3083, Australia

Abstract. Energy production in Australia is predominantly achieved with non-renewable sources such as coal and natural gas, with up to 86% of national energy being produced by these processes. Further, access to fresh water is fast becoming an issue in the harsh Australian climate. This paper proposes a novel design capable of producing electrical energy and desalinating water using only waste heat from industrial processes. A finned heat pipe heat exchanger plays an integral role in this process, extracting heat from a low-medium temperature waste heat stream and passing it through a thermoelectric generator and evaporative water desalination unit. Preliminary experimentation on this system showed that different desalination loop configurations played a large role in the desalination ability of the system, however had negligible effect on the thermal to electrical conversion efficiency of the system. When a closed loop system was used, the recovery ratio of the desalination unit reached 4% at the highest waste heat temperature. In the open loop configuration, the recovery ratio was below 1% for all tested inputs. The thermal to electrical conversion efficiency of this system was also measured and a value of about 1.2% was determined for most cases. Thus the ability of this novel system to both produce useful electrical energy and desalinate water was proven.

1. Introduction

Freshwater and useful energy are important for the growth of our society, particularly in Australia which produces most of its power through non-renewable methods and has problems with fresh water availability. This is a pressing issue in less developed regions of the country.

Thermoelectric generators (TEGs), in conjunction with desalination, provide an alternative hybrid approach to produce fresh water and power. Modern desalination processes primarily consist of multi-stage flash, multi-effect desalination and reverse osmosis. Multi-stage flash is responsible for 44% whilst reverse osmosis has risen to 42% of global desalination [1]. Multi-stage flash has specific energy consumption between 10-16 kWh/m^3, multi-effect desalination between 5.5-9 kWh/m^3 and reverse osmosis typically requires 3-4 kWh/m^3 with the inclusion of a recovery system [2]. In terms of the recovery ratio of each process, reverse osmosis tends to lead with ratios as high as 50%, followed by multi-effect desalination between 6-38% and multi-stage flash with less than 6% recovery ratio [3]. As of 2000, the combined world desalination capacity was 22 million cubic meters of water per day, requiring 203 million tons of oil annually as a fuel source [4]. Due to the vast energy requirements that conventional desalination requires, renewable sources typically are overlooked.

In 2012, 6.38 PJ of energy was rejected by the industrial sector as waste heat in the form of effluents and exhausts [5]. Recovery of even 1% of this waste heat is equivalent to over 3,000 tons of CO_2 from coal burning in Australia. The potential for energy recovery is thus clear within the industry.

* corresponding author: jvelardo12@gmail.com

Applications of TEGs have grown in the past decade. They have been used for recovery in automotive and industrial applications where the hot waste stream is used to generate electricity. The waste heat is absorbed by a cooling medium on the cold side of the TEG. Orr et al. [6] developed a TEG used for waste heat recovery in automobile exhaust streams. The TEG utilized heat pipes to help transfer heat from the hot waste stream (engine exhaust) to the cooling medium (which could be ambient atmospheric air). A total of 8 cells were used to create the TEG which was able to produce 6 W when charging the battery. The temperature of the waste stream was approximately 200°C at the inlet. This represented a conversion efficiency (thermal to electrical) of 1.43% which was 1/15 of Carnot efficiency at the same conditions.

Orr et al. [7] expanded on this work by introducing high temperature naphthalene heat pipes which only activated when dangerously high exhaust temperatures were experienced. The conversion efficiency in this case was 2.95% with higher waste heat temperatures than the previous work. A similar design of TEG waste heat recovery unit with heat pipe heat exchanger has been investigated by Remeli et al. [8]. Conversion efficiency of 0.7% was reported for this design. Champier conducted an investigation into the current state and application of TEG technology across all sectors, finding that typical TEG devices in use and current research exhibit a maximum efficiency of 5%, with the possibility of 10% being achieved in the coming years [9]. Despite their low efficiency, application toward waste heat recovery could still be beneficial. They are particularly useful when used in conjunction with heat pipes due to the large heat transfer rate associated with heat pipes. The cooling medium is seldom utilized for secondary, co-generative process such as evaporative desalination.

Research has been conducted into the combination of power generative systems with the added benefit of producing fresh water. Date et al. devised a system which utilized flashing of saline water to produce condensate and fresh water in a separate chamber [10]. They produced fresh water at a rate of 1.42 L/m and maximum power of 160 W with the use of a two-phase reaction turbine. Date et al., also proposed a novel waste heat recovery system with fresh water cogeneration. Heat pipe heat sinks and TEGs were used to produce 64 g of fresh water at a peak flux of 24 kW/m^2 [11].

In this work a waste heat recovery unit consisting of a heat pipe heat exchanger which was capable of both generating electrical energy and desalinating water was built and tested under various conditions. Such a novel design could help in combating the issues of industrial waste heat and limited access to fresh water.

2. Experimental setup

The proposed power generation and desalination system arrangement is shown in Figure 1. The waste heat stream here was simulated with hot air between the temperatures of 135-275°C. This is regarded as low to medium temperature waste heat and is generally neglected by the industry.

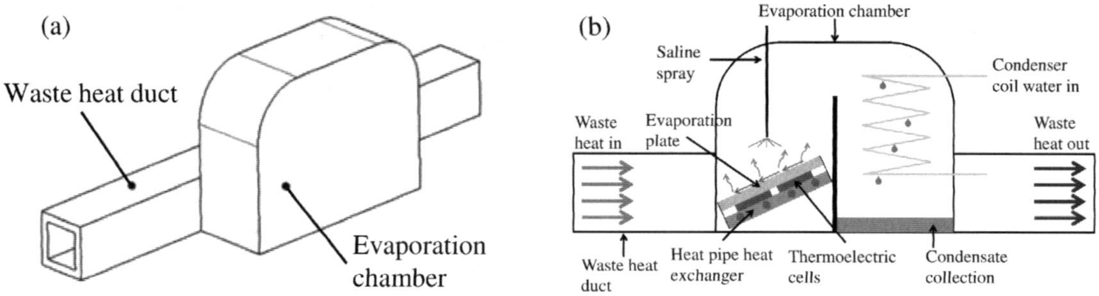

Figure 1. The basic arrangement of the combined waste heat recovery and water desalination system is shown in (a). The waste heat duct is located behind the evaporation chamber. Details of the arrangement are made clearer in (b).

The waste heat duct was well insulated however due to the high temperature of the waste heat it was unavoidable that some heat was lost to the atmosphere.

Integral to the design and operation of this setup was the heat pipe heat exchanger which transferred heat from the hot waste stream, through the TEG and into the saline solution. Four copper-water heat pipes were used in the heat exchanger as seen in Figure 2. These heat pipes had outside diameter and length of 8 mm and 300 mm, respectively. A copper fiber wick was used. The evaporator section of the heat pipes (exposed to waste stream flow) had total of 62 aluminium fins (153 mm x 50 mm x 0.5 mm) with fin spacing of 2 mm. This occupied the entire cross section of the waste heat duct to prevent air from flowing around the heat exchanger.

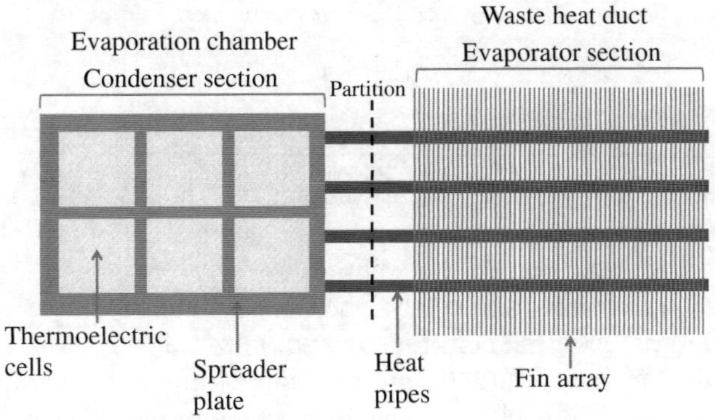

Figure 2. The heat pipe heat exchanger used in this study had four heat pipes. A fin array was used for the evaporator section (located in the waste heat duct) and the condenser section was embedded in a copper spreader block (located in the evaporator chamber).

The condenser sections of the heat pipes were embedded into a copper spreader block of size 100 mm x 150 mm x 15 mm (Figure 2). The condenser was located in a separate chamber to the evaporator (Figure 1). As shown in Figure 2, this copper spreader block also had an array of six thermoelectric cells on it, which were pressed between the copper spreader block and an aluminium plate (Figure 3). The thermoelectric cells were connected thermally in parallel, but electrically in series. Thermal paste was used to ensure good contact between the thermoelectric cells and the plates. The aluminium plate was maintained at a lower temperature than the waste heat stream by spraying a saline solution over it. The saltwater stream evaporated as it was heated by the aluminium plate and the vapour produced was condensed in a separate area inside the chamber, thus producing fresh water. A condenser coil with ambient temperature water was used to condense the vapour inside the chamber.

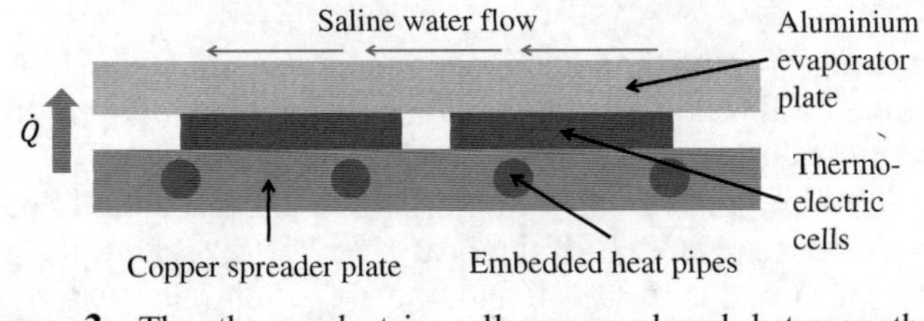

Figure 3. The thermoelectric cells were placed between the copper spreader plate of the heat pipe heat exchanger and the aluminium evaporation plate.

T-type thermocouples were present in several locations on the system to measure temperature. A hot-wire anemometer was used to measure velocity of air in the duct at various locations and an average air velocity was found for the duct. A variable resistor and multimeter were used to measure the electrical power generated by the TEG. Finally, the amount of fresh water produced was measured with a weighing scale. These measurements were then used to calculate system performance and efficiency, as shown in the following sections.

The tests were performed at three different average waste heat temperatures of 135, 215 and 275°C, representing low-medium temperature waste heat. A refractometer was used to establish the zero salinity datum of the saline solution used for all tests and then salt was added to a salinity of 1%.

Two different test configurations were trialed; an "open loop" and "closed loop" configuration. The closed loop configuration ensured that the saline solution flowed in a closed loop within the system i.e. once the saline solution flowed over the aluminium plate (where some evaporation had taken place) it would then be recycled to flow over the plate again. In the open loop configuration this was not the case; the saline solution would only pass through the system once.

3. Results

Firstly, the differences between the operating regimes (closed or open loop) will be discussed. It was observed through experimentation that steady state conditions were rarely achieved for the closed loop tests. This was due to the recycling of the saline solution. As the solution passed over the aluminium plate, heat was transferred to it from the waste heat stream (via the heat pipe heat exchanger). This would cause some sensible and latent heating of the solution as observed in the experimentation. Once the solution left the evaporation chamber (slightly warmer than it entered) it was then recycled back to the inlet. During this recycling process, the solution was not allowed to return to its original temperature and thus entered the system again at a slightly higher temperature than what it entered on the previous cycle. This meant that it was very difficult to achieve completely steady state conditions for the closed loop tests. Generally, this test had to be run for several hours until the temperature differences between components was reasonably steady with time, although the magnitudes of the temperature measurements still tended to increase. It was at this point that data measurements were taken and the analysis was conducted. For the open loop tests, steady state conditions were achieved in a much shorter period of time as discussed further below. In these open loop tests, the gradual heating of the saline solution wasn't an issue as the solution was only passed through the system once and not recycled.

Furthermore, regardless of the loop configuration, it was observed that increased waste heat temperature always lead to increased voltage and power generated by the TEG. This result is somewhat expected due to the general operating principles of thermoelectric cells.

For all tests, the freshwater was collected in the condenser section of the chamber. The salinity of the fresh water was measured by using a refractometer. The refractometer was initially used on fresh distilled water to gauge a datum value, and then salt was added to make the salinity of the solution 1%. After passing through the system, the salinity of the produced fresh water was equal to that of the distilled water before salt addition (according to the refractometer used). Thus the process used here was able to produce fresh water from a saline input.

3.1 Open loop tests

The steady state temperature profile of the 215°C open loop test has been shown in Figure 4. This is included to be representative of all open loop tests as the differences between these tests are only the magnitude of temperature, the trends were the same for all open loop tests.

For this particular test, it took approximately 35 minutes for steady state to be reached (not shown). It can be seen that the temperature upstream of the heat pipe heat exchanger is greater than that of the downstream temperature, indicating that heat is being transferred into the heat pipe heat exchanger from the waste stream. There exists a large temperature difference between the average hot and cold side of the TEG due to the large thermal resistance of the cells. The temperature of the TEG cold side was much larger than the cooling water in the condenser coil (not shown) which caused the migration of water vapour (evaporated from the saline solution) to the colder coil and the consequent condensation of vapour into liquid water, hence fulfilling the objective of water desalination from only a waste heat input. This is significant as many technologies require highly useful electrical energy as process inputs, whereas this process only required low-medium temperature (often neglected) waste heat as process input. This is an advantage of the current setup.

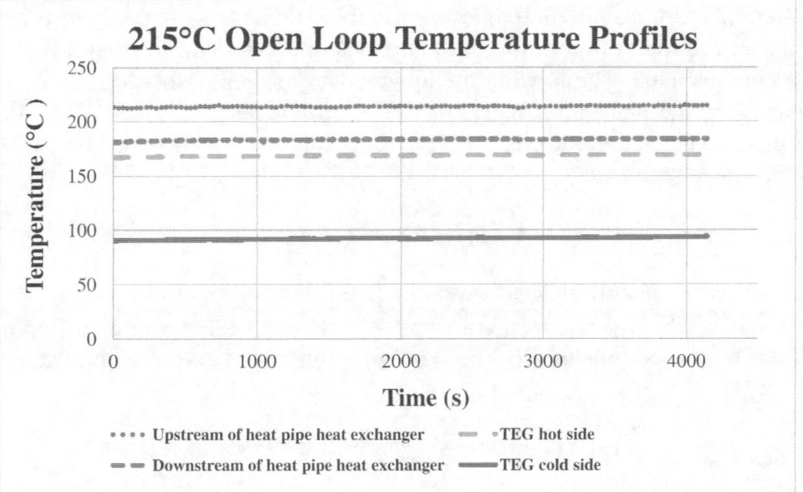

Figure 4. The temperature profile of the open loop test at waste heat temperature of 215°C. The trends are typical for all waste heat temperatures for open loop tests.

It can be seen that for this particular configuration, the temperature of the cold side of the TEG was slightly below 100°C, indicating that the majority of phase change of the saline solution was taking place due to evaporation. When the average temperature of the waste heat was 135°C, the TEG cold side was about 60°C indicating evaporation of the solution. At the highest average temperature of waste heat (275°C), the temperature of the cold plate reached 100°C, and flow boiling was beginning to be observed on the aluminium plate. Temperatures above 100°C could not be sustained for the cold plate in this test; it is thought that this was due to the flow boiling over the plate and the large latent heat of the saline solution. As the pressure inside the chamber was maintained at atmospheric, the boiling point of the solution was approximately 100°C (not exactly due to presence of salts). The effectiveness of boiling heat transfer processes ensure that the temperature of the cold plate never raised to a few degrees above 100°C.

In order to assess the effectiveness of the desalination process, the recovery ratio (*RR*) was defined as follows:

$$RR = \frac{m_{\text{desal}}}{m_{\text{feed}}} \tag{1}$$

Here m_{desal} is the total mass of the collected fresh water and m_{feed} is the total mass of the saline feed water. This was found to be a strong function of the waste heat temperature and generally increased as the temperature of the waste heat increased. The recovery ratio has been summarised in Table 1.

Table 1. Recovery ratio of open loop systems.

Waste heat temperature (°C)	Recovery ratio (%)
135	0.03
215	0.11
275	0.20

It was observed that the recovery ratio was generally quite small for all open loop tests. This can most likely be attributed to the open loop nature of the test, which meant that cool saline solution was continuously fed into the system and no recycling of the solution was allowed. The saline solution would thus enter the system at around ambient temperature and would have to sensibly heat before undergoing any latent heating. Due to the design of the flat plate evaporation surface, the solution was not allowed sufficient time to change phase which led to the low recovery ratios observed.

In order to assess the thermal performance of the system, measurements of the rate of heat transfer through the heat pipe heat sink and the electrical power output were recorded. The heat transfer rate was calculated from the waste stream as:

$$\dot{Q}_{\mathrm{th}} = \dot{m} c_p \Delta T \quad (2)$$

Here, $\dot{Q}_{\mathrm{th}}$ was the rate of heat transfer, $\dot{m}$ was the mass flow rate of the waste air stream, c_p was the average specific heat of the waste air stream and ΔT was the temperature difference between the upstream and downstream waste heat air. The thermal to electrical conversion efficiency, η, was then defined as:

$$\eta = \frac{\dot{W}_{\mathrm{elec}}}{\dot{Q}_{\mathrm{th}}} \quad (3)$$

Here, $\dot{W}_{\mathrm{elec}}$ is the electrical power output from the TEG. As can be seen in Table 2, large amounts of heat could be transferred through the heat pipe heat exchanger owing to the high effectiveness of the heat pipes. The conversion efficiency, defined as the ratio of electrical power output of the TEG to the heat transfer rate through the heat pipe heat exchanger was generally above 1%. It is unclear to the authors why the highest waste heat temperature test resulted in such low thermal efficiencies and relatively small heat transfer rate; it is believed there may have been a leakage in the system which led to erroneous measurement and poor thermal insulation of the system (consequently large heat losses).

Table 2. Thermal performance of open loop systems.

Waste heat temperature (°C)	Heat transfer rate (W)	Thermal to electrical conversion efficiency (%)
135	375	1.25
215	578	1.10
275	668	0.54

Conversion efficiencies of approximately 1% may not seem efficient, but this value is comparable to the values reported in the literature [6-9] for similar systems. This value highlights the potential of this setup. Improvements to the fin arrangement, thermal contact between components and better system insulation could increase this efficiency greatly. Furthermore, the Carnot efficiency at such low heat source temperatures is generally very low. For the lowest waste heat temperature of 135°C, the Carnot efficiency under similar conditions would be approximately 25%. The fraction of Carnot for this scenario is approaching 5%, which is a good result given the temperature of the waste heat.

3.2 Closed loop tests

For the closed loop tests, the temperature profile never completely reached steady state within the tested intervals for reasons discussed previously. Thus the temperature profiles have been omitted here however the trends seen in Figure 4 generally match those seen for the closed loop tests. Further differences were observed for the desalination and power generation capabilities of the system in open and closed loop configurations. Desalination will be discussed first.

As can be seen in Table 3, the recovery ratio for the closed loop systems tended to be larger than for the open loop systems. The reasons for this are simple; in the closed loop system the same saline solution is continually fed over the evaporation plate. Each pass of the evaporation plate causes the saline solution to sensibly heat by some amount. The whole solution is thus heated sensibly and brought up to boiling point temperature whereby phase change will take place. The differences are pronounced at waste heat temperatures of 215°C and 275°C as at these conditions the temperature of the evaporation plate tended to be near the boiling point of the solution, hence flow boiling was often observed and large amounts of solution could change phase and consequently be collected in the condenser section of the chamber.

Table 3. Recovery ratio of closed loop systems.

Waste heat temperature (°C)	Recovery ratio (%)
135	1.04
215	2.28
275	3.90

Clearly, the closed loop configuration is more appropriate from the point of view of desalination as recovery ratios as high as 4% were observed during testing. This is comparable with what can be achieved with a multi-stage flash system [3]. From the point of power generation output, the closed loop system is found to perform very similar to the open loop system. This is shown in Table 4.

Table 4. Thermal performance of closed loop systems.

Waste heat temperature (°C)	Heat transfer rate (W)	Thermal to electrical conversion efficiency (%)
135	353	1.19
215	473	1.19
275	735	0.98

It is obvious that for the closed loop test, the heat transfer rate tended to be lower than for the open loop tests (we are ignoring the 275°C test as the results of this test for the open loop system cannot currently be explained). This was most likely due to the always increasing saline solution temperature over the evaporation plate. The smaller driving temperature difference between the waste heat and saline solution for the closed loop system meant that less heat could be transferred through the heat pipe heat exchanger overall. In terms of thermal to electrical conversion efficiency, the results of both open and closed loop systems are somewhat similar and it is difficult to make judgement about which configuration provides better operation. It should be noted however that the voltage generated by the TEG was generally smaller for the closed loop tests than for the open loop tests. This was expected to be caused by the gradual heating of the TEG by the recycled saline solution bringing the temperature difference of the TEG sides closer together. This tends to lead to lower output voltages. It was interesting to note that the conversion efficiencies were still similar for both cases, even though output voltage produced were different.

Overall, the performance of the system tested here could use improvement to make it more viable, especially in regards to its desalination ability. Nevertheless the system showed its potential for combined power generation and water desalination. Recovery ratios as high as 4% were observed during testing, and electrical power outputs of 8 W were observed which represented TEG conversion efficiencies of about 1.2% (comparable to values reported in the literature for similar systems).

Integral to the design of this use is the use of heat pipes. The heat pipe heat exchanger was found to very effectively transfer large amounts of heat from the heat source (waste heat stream) to heat sink (desalination water). It was expected that replacement of the heat pipes with equivalent copper tubes would have led to much lower heat transfer rate at the same conditions, and also lower conversion efficiencies and desalination capabilities.

This system is expected to be improved in the future and the authors will report further on the developments. Given the results found from this experimental work, combined with the simplicity, compactness and multi-functionality of the system, further exploration is warranted.

4. Conclusions

In this work, a waste heat recovery unit capable of generating electrical energy and desalinating water was proposed and experimentally tested. Essential to the design of this system was a heat pipe heat exchanger which passed heat through thermoelectric cells (thus producing power) and into a saline solution causing evaporative desalination (in some cases flow boiling was observed). The experimental results showed promising system performance. The recovery ratio of this system reached as high as 4% and was a strong function of whether the system was operated in an open or closed loop configuration. Thermal to electrical conversions efficiencies of 1.20% were commonly observed and were not strongly dependent on the loop configuration.

Acknowledgements

The authors thank Fujikura Ltd. for providing the heat pipe heat exchanger used in this work.

References

[1] García-Rodríguez L 2003 Renewable energy applications in desalination: state of the art. *Solar energy*. Nov 1;**75**(5):381-93

[2] Ghaffour N, Missimer T M, Amy G L 2013 Technical review and evaluation of the economics of water desalination: current and future challenges for better water supply sustainability. *Desalination*. Jan 15;**309**:197-207

[3] Ali M T, Fath H E, Armstrong P R 2011 A comprehensive techno-economical review of indirect solar desalination. *Renewable and Sustainable Energy Reviews*. Oct 1;**15**(8):4187-99

[4] Kalogirou S A 2005 Seawater desalination using renewable energy sources. *Progress in energy and combustion science*. Jan 1;**31**(3):242-81

[5] Forman C, Muritala I K, Pardemann R, Meyer B 2016 Estimating the global waste heat potential. *Renewable and Sustainable Energy Reviews*. May 1;**57**:1568-79

[6] Orr B, Singh B, Tan L, Akbarzadeh A 2014 Electricity generation from an exhaust heat recovery system utilising thermoelectric cells and heat pipes. *Applied Thermal Engineering*. Dec 5;**73**(1):588-97

[7] Orr B G, Akbarzadeh A, Lappas P 2015 Reducing Automobile CO2 emissions with an exhaust heat recovery system utilising thermoelectric generators and heat pipes. *SAE Technical Paper*; Mar 10

[8] Remeli M F, Date A, Orr B, Ding L C, Singh B, Affandi N D, Akbarzadeh 2016 A Experimental investigation of combined heat recovery and power generation using a heat pipe assisted thermoelectric generator system. *Energy conversion and management*. Mar 1;**111**:147-57

[9] Champier D 2017 Thermoelectric generators: A review of applications. *Energy Conversion and Management*. May 15;**140**:167-81

[10] Date A, Khaghani A, Andrews J, Akbarzadeh 2015 A Performance of a rotating two-phase turbine for combined power generation and desalination. *Applied Thermal Engineering*. Feb 5;**76**:9-17

[11] Date A, Gauci L, Chan R, Date A 2015 Performance review of a novel combined thermoelectric power generation and water desalination system. *Renewable energy*. Nov 1;**83**:256-69

Loop thermosyphons with porous coatings

L L Vasiliev*, M I Rabetsky, L P Grakovich, V K Kulikouski, A S Zhuravlyov and M A Kuzmich

A V Luikov Heat and Mass Transfer Institute, National Academy of Sciences of Belarus, 15 P. Brovka Str., Minsk, 220072, Belarus

Abstract. The aim of this work is to experimentally evaluate the heat exchange of evaporation and condensation inside a new annular thermosyphon with horizontally located zones of evaporation and condensation, which are interconnected by a vapor line and a condensate line. The mini-tube of the liquid line is introduced coaxially into the cylindrical evaporator with a porous coating, thus an annular channel is formed between the tube and the porous coating of the evaporator. The purpose of this study is to determine the thermal resistance of the device and its components, as well as the temperature field along the thermosyphon. The considered thermosyphon guarantees a shorter start-up time, lowers the temperature of the evaporator walls, has a small temperature hysteresis with an increase/decrease in the transferred heat flux, and suppresses temperature pulsations inside an evaporator. The 130 mm long evaporator is made of copper tube (d_{out} = 12 mm, d_{in} = 10 mm). The wick is obtained by sintering copper powder. The working fluids R245fa and R600 were used. The thermal resistance of thermosyphon does not exceed 0.5 K/W, and of the evaporator 0.1–0.05 K/W).

1. Introduction

Heat pipes and similar two-phase thermosyphons are efficient, compact and passive heat transfer devices. The increase in the productivity of equipment, which is accompanied by an increase in the intensity of heat release (electronics, telecommunication systems, solar technology with concentrators of incident solar radiation) stimulates the creation and production of new devices that meet modern requirements for thermal control of equipment. The current trend in the development of equipment for micro- and optoelectronics (MEMS) is to improve its layout, reduce the assembly due to the miniaturization of units in connection with needs of their commercial applications. Loop heat pipes (LHP), loop thermosyphons (LT), mechanically pumped two-phase loops, and evaporative spray cooling, pulsating heat pipes play a vital role in heat dissipation and recently have been widely used in the electronic components cooling and some other energy application [1–7]. Due to the aptness to transfer large heat fluxes over long length without the use of mechanical energy, as well as reliability, high resource, technical simplicity and low cost, thermal management systems based on the above devices are very suitable. The main advantage of such models is a separation of vapor and liquid flows, they move through different pipes. This technology guarantees a higher heat transfer capacity of heat pipes and thermosyphons.

*Corresponding author: leonard_vasiliev@rambler.ru

Recently a new polymer loop thermosyphon was suggested and tested [8]. The horizontally arranged flat evaporator and condenser are coupled by a flexible vapor line and a condensate line made of polymer material.

In [5], a study of a model of a two-phase loop with three evaporators is described; the movement of the coolant is carried out by forced convection. The purpose of the experiments was to determine the performance of a system with parallel operating evaporators at different levels of heat load on each of them. The applied porous evaporation surfaces made it possible to increase the heat transfer coefficients up to 20 $W/(cm^2 \cdot K)$. The results of the study of a loop thermosyphon with several evaporators at various values of heat load and angles of inclination are presented in [6]. The device is a closed planar serpentine made of an aluminum tube with an inner diameter of 3 mm; FC-72 fluid was used as a working fluid. The authors of [7] studied both numerically and experimentally the effect of the angle of inclination and the degree of wettability of the evaporator surface on the heat transfer capacity of the thermosyphon.

Cryogenic loop heat pipes are of great interest for cooling and thermal regulation at low temperatures (some medical and infrared sensors, superconductive magnetic materials) [9, 10]. A low-temperature medical heat pipe probe for non-invasive, non-drug treatment by hypothermia of a local human cavity was suggested and tested in [11]. Heat transfer processes in heat pipes and two-phase thermosyphons in combination with various technologies including nanocoatings, nanofluids, and nanocomposites are analyzed in detail in some publications [12–16]. Modern loop heat pipes for the electronic components cooling with sophisticated types of porous membrane were designed and tested in [17–22]. For use in these experimental programs, three evaporators with different modifications of wicks, such as microchannel, modulated monoporous, and modulated biporous wicks, were manufactured and combined with other constituent elements of the LHP. Various types of thermosyphons have been developed and studied for use in power systems [22], nuclear reactors [23, 24], solar heat recuperators [25–27], air conditioning [28] cooling systems for microelectronics [29], electronic components in avionics [30], and other areas.

The devices mentioned effectively work under optimal conditions. For thermosyphons this is a vertical or close to vertical position, heat pipes are less sensitive to orientation in space, but their heat transfer capacity drops at low hydrostatic pressures. Meanwhile, in some cases, when solving problems of heat removal, it is required to locate an evaporator and condenser horizontally, while from the layout conditions they should be quite long. Loop heat pipes [31] are capable of transferring heat flow in a horizontal direction and even in a direction opposite to the gravity vector, however, loop heat pipe has limitations on the length of the evaporator, and there may be some problems with starting, such as a long start-up time or thermal and hydraulic oscillations at low heat loads. The specified reasons were the motivation for the development and study of a new design of thermosyphon. This study is oriented on the tests of the loop thermosyphon with porous coating of the evaporator (LTCE). This is the new device, the main difference of it from traditional thermosyphons and loop heat pipes is the ability to transfer heat flow in a horizontal direction, while the condenser can be extended compared to loop heat pipe, this expands the possibilities of using a thermosyphon. The simplest LTCE consists of an evaporator with a compensation chamber and a condenser, which are interconnected by vapor and liquid pipes. The proposed concept is to use a new combined two-phase mode of pumping the working fluid through the mini-gap of the annular channel in the evaporator. Due to its unique characteristics (high thermal conductivity, extended, evaporator, etc.), a new loop thermosyphon with a porous evaporator is proposed for cooling objects in electronics, in medicine as a "cold finger" for cooling the human body, in solar photovoltaic and heating systems. Cryogenic infrared sensors, based on the application of LTCE coolers ensure precise thermal control at a low temperature. LTCE could solve the problem of cooling down superconductive magnetic materials directly or indirectly. It is interesting to note that the start-up of the LTCE has some advantages over the conventional LHP and LT (the transport distance and heat release capacity). The LTCE as an advanced electronic cooling device possesses the ability to handle multiple heat sources (chips, LED emitters, et.) simultaneously located along the single evaporator surface.

Based on the above mentioned problems, this study focuses on carrying out the work on the developing, designing and testing LTCE and, as a result, obtaining a solution that can significantly improve its efficiency. The results show that the thermosyphon is capable of transferring heat load of 140 W at a heat flux of 18.9 W/cm^2 in an evaporator. The total thermal resistance of the device is 0.36 K/W.

2. Loop thermosyphon with the evaporator having a porous coating

The copper LTCE (Figure 1) is equipped with an extended (length 100–200 mm) horizontal cylindrical evaporator, 10–12 mm in diameter, and a cylindrical condenser, which was cooled using a flow-through liquid heat exchanger in the described experiments. The inner surface of an evaporator is covered with a porous layer of no more than 1 mm thickness of sintered copper powder particles. When the volume of a compensation chamber is correctly selected and the filling ratio is optimal, a liquid supply is constantly present in a compensation chamber, and the LTCE shows an ability to self-regulate an amount of liquid in an evaporator.

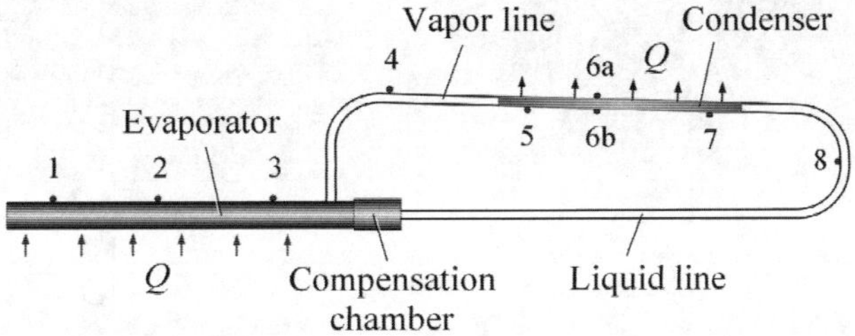

Figure1. LTCE scheme. Points 1–8 indicate the positions of thermocouples.

The LTCE evaporator (Figure 2) has some important design features, including an annular gap. A porous wick (porosity ~ 50%, average hydraulic pore diameter 0.025 mm) was made by sintering copper particles of 0.063–0.100 mm fraction (average diameter 0.082 mm). During a heat load application to an evaporator, intense heat and mass transfer occurs in a narrow annular channel. The evaporator can be divided into four conventional zones with the following heat transfer modes: subcooled flow boiling, flow nucleate boiling, and evaporation of the working fluid in an annular gap and a porous wick. The movement of the liquid phase is caused by a pressure drop due to the force of gravity, a formation and growth of vapor bubbles, and capillary forces in a porous media.

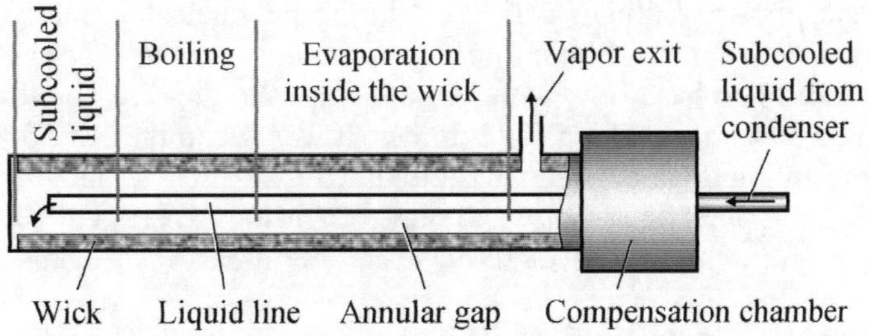

Figure 2. Longitudinal section of the LTCE evaporator and zones of hydrodynamic and heat transfer modes disposed along them.

Before the decision was made concerning the choice of the best type of porous structure and the LTCE evaporator final design, some experiments were carried out on an experimental setup, the main element of which is a boiling vessel placed in a hermetic heat-insulated box equipped with a system for maintaining the required temperature to reduce a heat exchange between the vessel and the environment (a refrigeration machine, a thermostat, a developed channel circuit on the inner walls and a pump for circulation liquid of a given temperature). The boiling vessel has a working section with a test sample and an electric heater, a condenser, windows for observing and photographing the process, a sleeve for introducing thermocouples to control the temperature. Two sets of experiments were realized to study a heat transfer and hydrodynamics during the phase transition on the surface of the test specimen. The first set of experiments was conducted in the liquid pool (macroscale heat transfer). The second set was realized inside the system of mini-/microchannels (porous material). A more detailed description of the equipment and the method of experiments are presented in [20]. In our experiments, a sample (a horizontal tube 20 mm in diameter) was coaxially placed in a transparent glass cylinder with an annular mini-channel 0.8–1.8 mm (Figure 3).

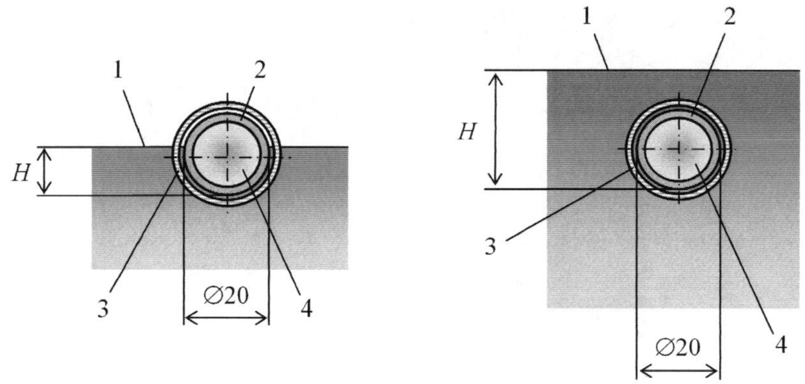

Figure 3. The sectional drawing of the tested sample partially sumberged (left) and put in the liquid pool (right): 1) liquid surface, 2) test sample, 3) glass tube, 4) electrical heater, H is the liquid column height.

All measurements were carried out in a stationary mode of the process. The heat transfer coefficient h was determined as

$$h = q/(T_w - T_l), \quad (1)$$

where q is a heat flux; T_w and T_l are the temperatures of wall (heat exchange surface) and liquid.

Visual observation of two-phase heat transfer testifies that movement of vapor bubbles in an annular mini channel is complex. Bubbles generated during the nucleate boiling can be considered as micro pumps action to move vapor and liquid both perpendicular to the heat exchange surface and to create mini-pulsations of a two-phase flow along a sample axis in the direction from an entrance to an exit to a liquid pool in an upper zone of a glass cylinder. It was found that a heat transfer coefficient is affected by a position of an interface between a liquid and a vapor space in a glass cylinder: it took on maximum values when a sample was partially submerged. During experiments, it was determined that a number of menisci in pores and, consequently, a surface area of an evaporation increases with increasing heat flux.

The process modeling an evaporator of LTCE is shown in Figure 4: 1) heating of subcooled liquid zone (Fig. 4a); 2) transient zone of subcooled liquid boiling in the mini channel, Fig. 4b; 3) zone of the developed nucleate boiling in the mini channel (Figure 4c); 4) liquid rivulet at the bottom part of glass tube and evaporation inside the porous wick (Figure 4d).

Figure 4. Modes of heat transfer in an evaporator: a) subcooled liquid heating near the entrance of the annular mini channel of LTCE evaporator; b) beginning of nucleate boiling of the saturated liquid in the annular mini channel of LTCE evaporator; c) bubble flow boiling heat transfer; d) evaporation inside the wick, capillary suction of liquid from the liquid rivulet at the bottom of evaporator.

The experimental data obtained on submerged and partially submerged porous tube are shown in Figure 5. In a certain range of heat loads, the intensity of heat transfer in a narrow annular channel is 2.5–3 times higher than during pool boiling.

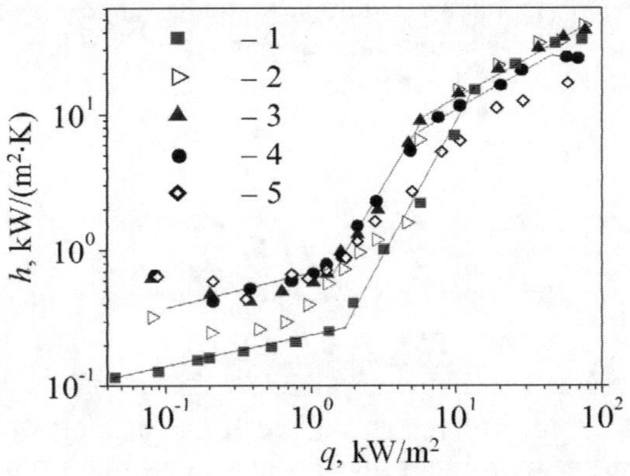

Figure 5. Heat transfer intensity along the evaporator depending on a height H of a liquid inside the annular mini gap: 1–5 – H = 70, 2, 1.5, 1, and 0.5 mm.

Open porosity, defined as the ratio of a pore volume that can be filled with liquid to a total volume of a body, plays an important role in a development of hydraulic and thermal processes during phase transitions and, therefore, is one of the key characteristics of a porous wick. Featuring a set of mini- and macro-scale pores, the wick provides a high mass flow rate for fluid filtration. Another essential parameter of a wick is permeability, which affects a rate of fluid flow through pores.

Various processes occur in the annular mini-channel of the evaporator, including capillary pumping of the liquid phase through the pores, two-phase heat transfer in the form of evaporation and boiling in the flow. In the case under consideration, only walls of an evaporator have a porous coating, while channels for vapor and liquid are made of smooth-walled tubes, which significantly reduces a hydraulic resistance and, as a result, a pressure loss inside them. A liquid line connecting the LTCE condenser and evaporator is inserted into an evaporator in such a way that an annular mini gap is formed between its extension inside an evaporator and a porous wall of an evaporator (Figure 2), inside of which it provides forced nuclear boiling and evaporation with intense heat transfer.

The LTCE condenser is made as a straight copper tube 105 mm long, 6 mm diameter disposed inside the liquid/liquid heat exchanger. A condenser envelope is cooled with water transferred from the tank through a flow-through heat exchanger that encloses a condenser. The temperature of cooling water is kept stably by means of the thermostat; the volumetric flow is controlled by a flow meter. The thermal control system allows regulating the temperature of the cooling water in the range of 20-80 ^{0}C. Thus there is a possibility accordingly to change the temperature of vapor condensation, changing the boundary conditions of heat transfer inside the LTCE condenser. A total length of the LTCE tracts, including vapor and liquid lines, is about 500 mm.

The pressure balance during a circulation of a vapor and liquid phases of a working fluid in the LTCE circuit is expressed as follows:

$$\Delta P_g + \Delta P_{bg} = \Delta P_l + \Delta P_v + \Delta P_e + \Delta P_c, \tag{2}$$

where ΔP_g is the pressure drop due to gravitational forces, ΔP_{bg} is the pressure drop due to the formation and growth of vapor bubbles in an annular channel; ΔP_v and ΔP_l are pressure drops in vapor and liquid flows; ΔP_e and ΔP_c are pressure drops due to phase transitions (evaporation and condensation). To ensure a proper functioning of the LTCE, it is necessary that the sum of $P_g + \Delta P_{bg}$ must be greater than or at least equal to the sum of all terms on the right side of equation (2). There is also a capillary pressure drop ΔP_{cap} along the porous coating perimeter; ΔP_{cap} is responsible for a liquid capillary suction inside a porous coating from a liquid rivulet inside an annular channel, but does not significantly affect working fluid circulation along the LTCE.

The average difference ΔT_{avg} between the temperatures of an evaporator and a condenser when calculating according to Eq. (3) was determined using the surface-averaged temperatures of a condenser T_c and the evaporator T_e, which were calculated using Eqs. (4) and (5):

$$\Delta T_{avg} = T_e - T_c, \tag{3}$$

$$T_e = (T_1 + T_2 + T_3)/3, \tag{4}$$

$$T_c = (T_5 + T_6 + T_7)/3, \tag{5}$$

where T_1–T_3 and T_5–T_7 are the temperatures on surfaces of an evaporator and a condenser, T_4 and T_8 are the temperatures on a vapor and a liquid lines. Numbers 1–8 indicate positions of thermocouples (Fig. 1). The LTCE with an evaporator with a diameter of 10–12 mm and a length of up to 200 mm is able to prevent heating of a cooled element above 80°C at a heat load of 200 W. The LTCE design provides quick start-up, lowering an evaporator shell temperature, and stabilizing a temperature inside a thermosyphon, suppressing its changes. One of the important properties of LTCE is an aptitude to transport heat a long distance from a heat source. Heat transfer in LTCE is carried out by the action of gravity forces and pressure impulses resulting from nucleate boiling of condensate in a mini-channel of an evaporator.

According to published data on studies of loop heat pipes [11, 21], when testing copper-water LHPs with wicks of copper or less thermally conductive materials, in order to guarantee a successful start-up of a device, a vapor temperature should not be lower than 40°C even with minimal heat loads due to low water pressure at room temperature.

A thermal resistance of LTCE decreases with increase in a temperature of a liquid used for heat removal from the condenser (liquid line). A condenser/radiator is a high-mass element with a developed surface. The rate at which liquid accumulates in a condenser is determined by a number of factors:

- temperature of a sink;
- heat load affecting a vapor generation rate;
- absolute temperature and pressure values;
- fluid thermophysical properties;
- thermal conduction in a face sheets;
- back thermal conduction;
- vapor entering into an evaporator through a condenser;
- degree of subcooling of a liquid entering an evaporator due to accumulation of liquid and changes in the temperature of a radiator.

The ability to use a "hot" cooling liquid has an important practical prospect in particular for "free cooling" technology, for example, in cooling systems of data-centers, utilization of heat dissipated inside data-centers. The cooling capacity of the LTCE condenser (heat sink) is determined by the following formula

$$Q_{out} = m \cdot c_p (T_{out} - T_{in}), \tag{6}$$

where m, c_p, T_{out}, and T_{in} are respectively the flow rate, specific heat at a constant pressure, the outlet and inlet temperature of cooling water.

3. Experiments

Many factors such as optimal geometric parameters, liquid filling ratio, heat load, tilt angle affect, and the start-up time of the LTCE were considered in the experiments. In order to establish the heat transfer capacity of the LTCE, a set of experiments was carried out using various working fluids. The diagram of the experimental setup is presented in Figure 6.

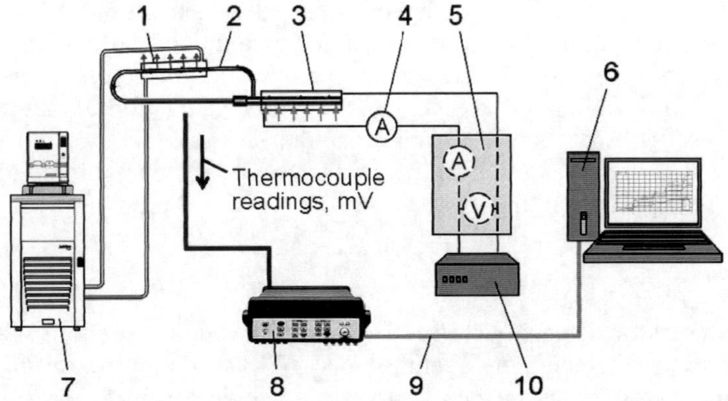

Figure 6. Scheme of the experimental set-up for LTCE testing: 1) liquid flow heat exchanger, 2) LTCE, 3) electrical heater, 4) ammeter, 5) wattmeter, 6) computer, 7) thermostat, 8) Agilent for data acquisition/logger for data registration, 9) signal cable for thermocouples, 10) regulated power supply.

An electric heater was applied as a source of energy. Heat flow Q, defined as the electric power supplied to the heater, taking into account heat losses, was absorbed by the water cooling system of the LTCE condenser. The liquid temperature inside the Circulating Thermostatic Bath Carbolite Gero Limited, Julabo FP-89 was changed from the "cold" to "hot" temperature (20 to 80°C) with a liquid flow rate of 10 L/min. The precision of the liquid temperature maintenance in the Bath is ±0.1°C. Furthermore, the measured data were recorded by the Agilent 34970A data acquisition system, which was subsequently processed. To measure a working vapor pressure inside the LTCE during the experiments, a PSA-C01 pressure meter, and an NI-9203 were used. A temperature field along the LTCE was registered by thermocouples placed on a surface of an evaporator and condenser envelops (Fig. 1). The measurement errors were determined in accordance with the methodology proposed in [32]. The precision of the heat flow measurements was ±0.5%; temperature uncertainty was about ±1.25 °C. To minimize heat transfer with the environment, the system was reliably insulated, thus enabling a correct assessment of the thermal performance of the LTCE. LTCE samples were tested using two working fluids, R245fa and R600. In these experiments, a heat load applied in an evaporator starting from 10 W was increased up to 30, 40, 50, 60, 70, 80, 90, 100, and 120 W and then was decreased back. No hysteresis of LTCE thermal resistance via heat load decreasing was observed.

The thermal resistances of an evaporator R_e, a condenser R_c, and total R_t for the LTCE are calculated as follows:

$$R_e = (T_e - T_v)/Q, \quad (7)$$

$$R_c = (T_v - T_c)/Q, \quad (8)$$

$$R_t = (T_e - T_c)/Q, \quad (9)$$

where T_e and T_c are temperatures of an evaporator and a condenser, R_t is a total thermal resistance of LTCE.

4. Results and discussion

In the present research, a cycle of experiments was conducted to study and analyze some LTCE parameters, including the thermosyphon geometry, liquid filling ratio, vapor and liquid line lengths, and heat load. It is known that a functioning of a closed system in which combined process of evaporation/boiling and condensation occur accompanied by a mobility of an interface between vapor and liquid, which is in principle impossible without temperature and pressure pulsations. The same phenomena are typical for LTCE. A difference may lie in an organization of mentioned processes and the values of pulsations. If a process of heat and mass transfer in LTCE is organized optimally, using evaporation in a porous structure as the main heat transfer mode, the amplitude of pulsations will be very small and it is of little practical importance.

4.1. Results

4.1.1. Copper-R245fa LTCE #1.

The geometric parameters of Copper-R245fa LTCE thermosyphon, sample #1 are presented in the table 1. The start-up and the temperature evolution of LTCE #1 via a time for different heat input Q are presented in Figure 7.a. The temperature curves are shown for an evaporator and a condenser, working fluid in vapor and liquid pipes, and inside a condenser (heat sink). The obtained results show that LTCE (copper-R245fa) has a good dynamic of a start-up. It starts at sufficiently low heat loads (less than 10 W). The LTCE is stable in operation over a wide range of heat input (10–100 W). The total thermal resistance R_t is 0.34 K/W. The evaporator thermal resistance R_e is equal 0.1–0.05 K/W in a heat load range 30–100 W.

Table1. Dimensions of the LTCE components (sample #1)

Characteristics		Size, mm
Evaporator	Length	130
	Diameter/wall thickness	12/1
Condenser	Length	105
	Diameter/wall thickness	4/0.5
Vapor line	Length	90
	Diameter/wall thickness	4/0.5
Liquid line	Length	350
	Diameter/wall thickness	3/0.5

The temperature graphs along the LTCE are given in Figure 7.b. The experimental data indicate that the temperature drop along a length of an evaporator is small. When a thermal load of the LTCE evaporator is changing from 30 up to 120 W, a heat transfer capacity of the thermosyphon is high and not deteriorated.

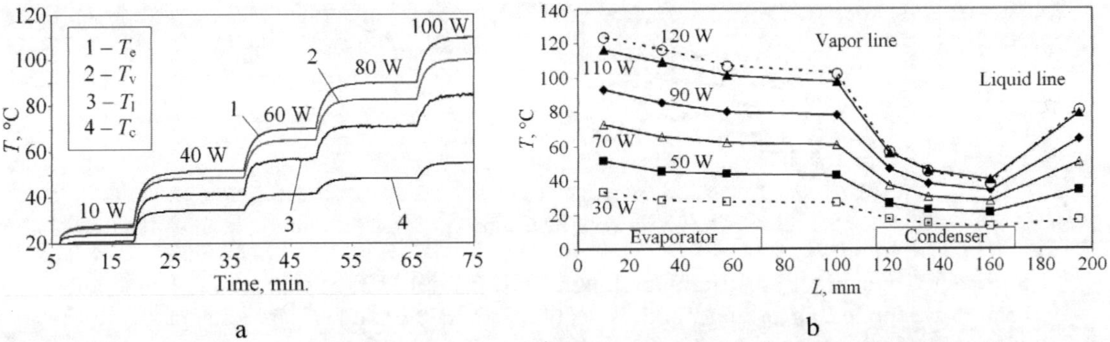

Figure 7. (a) Start-up of LTCE, sample #1. T_e, T_v, T_c, and T_l are the temperature of evaporator, vapor, condenser, and liquid respectively. Working fluid – R245fa (b) Temperature distribution along LTCE, sample #1, as a function of heat load in an evaporator. Working fluid – R245fa.

The thermal resistance of an evaporator, condenser and thermosyphon LTCE as a function of temperature of a cooling liquid is shown in Figures 8a and 8b. A heat load is equal 40 W. It should be noted that with increase of the temperature of a coolant liquid in a thermostatic bath the total thermal resistance LTCE decreases, Figure 8.b. This is very important, since the temperature of the working liquid entering the evaporator is increasing and the boiling of the saturated liquid is beginning earlier. For example the air temperature inside a closed space like a data center can be high to compare with a surrounding.

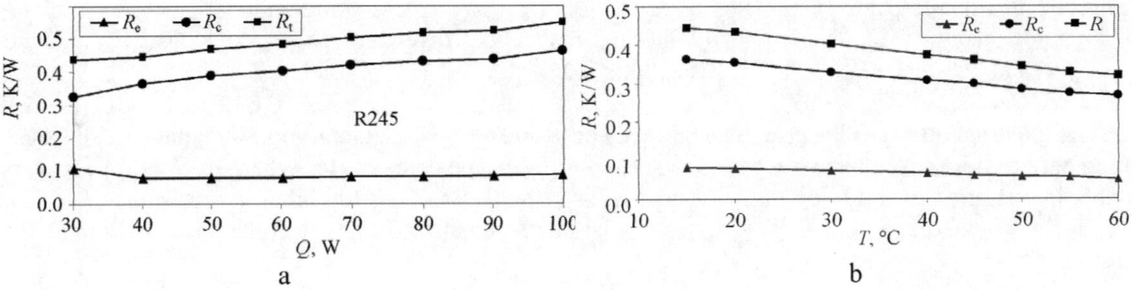

Figure 8. (a) Thermal resistances of evaporator R_e and condenser R_c and a total thermal resistance R_t of LTCE #1 as a function of the LTCE heat load (b) Thermal resistances of evaporator R_e and

condenser R_c and a total thermal resistance R_t of LTCE #1 as a function of a temperature of a cooling water, $Q = 40$ W.

4.1.2. Copper-R600 LTCE #2.

The dimensions of the LTCE#2 thermosyphon charged with R600 are the same as those of LTCE#1 (table 1). A temperature evolution of the LTCE #2 components via a time and along the LTCE for different heat input is shown in Figures 9 and 10a. The thermophysical parameters of the LTCE sample #2 with the working fluid R600 are similar to LTCE thermosyphon, sample #1. The thermal resistances of the evaporator, condenser, and thermosyphon as a function of the temperature of the cooling liquid are shown in Figure 10b, depending on the heat load of LTCE.

The thermal resistance of LTCE decreases with the temperature increase of the coolant liquid.

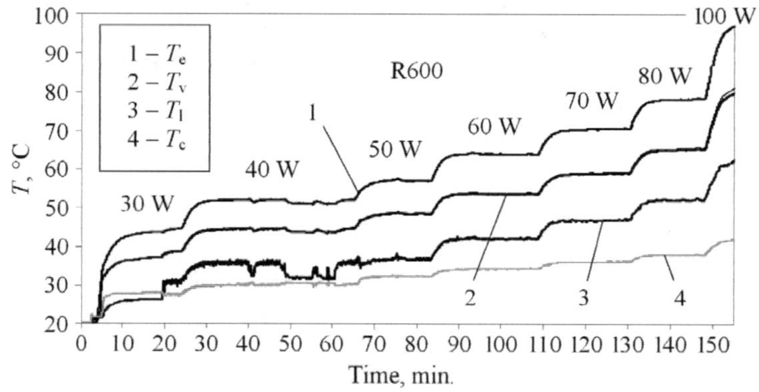

Figure 9. Temperature distribution along LTCE (sample #2) as a function of time and heat flow applied. Working fluid is R600. T_e, T_v, T_c, and T_l are the temperature evolution of an evaporator, a vapor, a condenser, and a liquid, respectively.

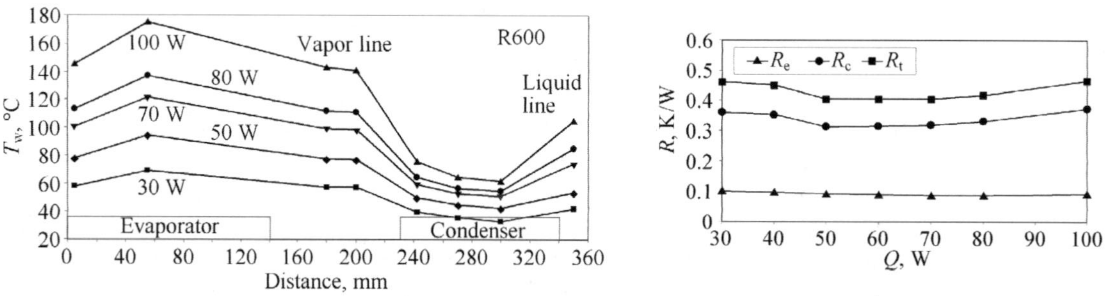

a b

Figure 10. (a) Temperature profiles along LTCE (sample #2) as a function of heat load in the evaporator. Working fluid is R600 (b) Thermal resistances of evaporator R_e, condenser R_c and the total thermal resistance R_t of LTCE #2 as a function of the LTCE heat load.

The experimental results show that R245fa and R600 LTCE's have a good dynamics of the start-up. Thermosyphons are easily starting at sufficiently low heat loads (less than 10 W) and are very stable in operation over a wide range of heat loads (10–100 W). The total thermal resistance R_t of the LTCE does not exceed 0.35–0.5 K/W (for evaporator R_e = 0.1–0.05 K/W) in the heat load range 30–100 W.

4.2. Discussion

LTCE thermosyphons summarize the advantages of conventional heat pipes, LHPs and LTs. The experimental results show that the modified system of LTCE is able to prevent the unacceptable surface drying phenomena and significantly reduce the evaporator temperature.

According to some authors [16, 21], the start-up at low heat loads is a problem operating stage for conventional LHPs. This problem is not important for the case of LTCE. Even at low heat inputs a easy starting of LTCE takes place. There is good circulation of the working fluid is ensured inside LTCE thermosyphon in the wide interval of heat loads. The porous coating of the evaporator minimizes fluctuations of temperature, pressure and the noise. The application of porous coating in the LTCE evaporator increases the two-phase heat transfer intensity nearly 2–3 times than at a pool boiling heat transfer in a conventional loop thermosyphon with a smooth-wall evaporator.

5. Conclusions

The paper describes a new looped thermosyphon (LTEC) concept, in particular a bayonet-type annular evaporator with a porous coating to facilitate enhanced boiling. The thermosyphon configuration is an evaporator (heat supply zone) with a compensation chamber and a liquid-cooled condenser (heat sink), connected by a vapor line and a liquid (condensate) line, Figure 1. Two experimental thermosyphons LTEC were made of copper tube, the geometric dimensions are given in table 1. Both models of LTCE have the same wick, which is made of copper sintered powder. The working fluids used are R245fa and R600. The LTCE model #1 with R245fa is able to withstand heat loads up to 130 W, providing evaporator heating not higher than 50°C. The LTCE concept is very promising for efficient thermal management of cooling data centers, servers, and other electronic devices operating under high thermal loads. As a result of tests, it was found that the thermal resistance of LTCE decreases with an increase in a supplied heat flux, which is explained by an increase in a circulation rate of a working fluid. Due to passive phase separation, there is no need for active flow control even with very short local increases in heat flow along the evaporator. It is very significant that, unlike a loop heat pipe, the LTCE evaporator is practically unlimited in length. Consequently, a number of cooled elements, such as cryogenic infrared sensors, electronic chips, LEDs, etc., can be attached to one unique evaporator of LTCE. This device has good dynamic characteristics: it starts up at sufficiently low heat loads (less than 10 W), quickly goes into operation and works stably in a wide range of heat loads. In this case, the thermal resistance of the LTCE evaporator is 0.05–0.1 K/W for both tested fluids. The heat flow density for LTCE #2 (L = 130 mm) is equal to q = 48 kW/m^2. At a heat load of 100 W, the surface temperature of the LTCE evaporator is 105 C for R245fa, and near 100°C for R600.

6. Conclusions

c	– specific heat, J/ (kg K)
d	– diameter, mm
H	– height, mm
h	– heat transfer coefficient, W/ (m^2 K)
L	– length, mm
m	– mass, kg
P	– pressure (Pa)
Q	– heat flow, W
q	– heat flux, W/m^2
R	– thermal resistance, K/W
T	– temperature, °C

Subscripts:

avg – average;
bg – bubble generation;
c – condenser, condensation;
cap – capillary;

e – evaporator, evaporation;
g – gravitation;
in – input, inner;
l – liquid;
out – output, outer;
t – total;
v – vapor;
w – wall (heat exchange surface);

Acronyms:
LHP – loop heat pipe,
LT – loop thermosyphon,
LTCE – loop thermosyphon with porous coating of the evaporator

References

[1] Garner S and Patel C 2001 *Proc. The Pacific Rim/Int., intersociety electronic packaging technical/business Conf. (Kauai, Hawaii, USA)* in: *Advanced Electronic packaging* (ASME) pp 961–968 IPACK2001-15782
[2] Singh R, Akbarzadeh A and Mochizuki M 2008 Operational characteristics of a miniature loop heat pipe with flat evaporator *Int. J. Therm. Sci.* **47** 11 pp 1504–1515. https://doi.org/10.1016/j.ijthermalsci.2007.12.013
[3] Ranasamy N S, Kumar P, Wangaskar B, Khandekar S and Maydanik Y F 2018 Miniature ammonia loop heat pipe for terrestrial applications: Experiments and modeling *Int. J. Therm. Sci.* **124** pp 263–278. https://doi.org/10.1016/j.ijthermalsci.2017.10.018
[4] Khrustalev D 2002 Proc. 18th SEMI-THERM Symp. *(San Diego, California, USA)* pp 145–150
[5] Winter M and Stephan P 2012 Evaporation from micro-porous surfaces in mechanically pumped two-phase loop *Proc. 16th Int. Heat Pipe Conf. (*Lyon, Fr*)* pp. 295–300
[6] Mamelli M, Mangini D, Vanoli G F , Araneo L, Filippeschi S and Marengo M 2015 Multi-evaporator closed loop thermosyphon *7th Eur.-Jpn. Two-Phase Flow Group Meet. (Zermatt, Switz)* pp 1–12
[7] Zhi Hu, Yaning Zhang, Bingxi Li, Chi-Chuan Wang and Yongji Li 2018 The influence of the inclination angle and an evaporator wettability on the heat performance of the thermosyphon by simulation and experiment *Int. J. Heat Mass Transf* **116** pp 675–684
[8] Vasiliev L L, Grakovich L P, Rabetsky M I and Vassiliev L L, Jr 2013 Heat transfer enhancement in heat pipes and thermosyphons using nanotechnologies (nanofluids, nanocoating, and nanocomposites) as an HP envelope *Heat Pipe Sci. and Technol. Int. J.* **4** 4 pp 251–275
[9] Zhao Y, Yan T and Liang J 2011 Experimental study on a cryogenic loop heat pipe with high heat capacity *Int. J. Heat and Mass Transf.* **54** pp 3304–3308
[10] Khrustalev D Cryogenic loop heat pipes as flexible thermal links for cryocoolers 2003 *Cryocoolers 12* ed R G Ross Jr (New York: Kluwer Academic/Plenum Publisher) pp. 709–716; 12:709-16, https://doi:10,1007/0-306-47919-2_93
[11] Vasiliev L L, Zhuravlyov A S, Molodkin F F, Khrolenok V V., Zhdanov V L., Vasil'ev V L., Adamov S I and Tulin A A 1996 Medical instrument based on a heat pipe for local cavity hypothermia *J of Eng. Phys. and Thermophys.* **69** 3 pp 302–304
[12] Vasiliev L L., Vassiliev L L, Jr, Zhuravlyov A S, Shapovalov A V and Rodin A V 2015 Vapordynamic thermosyphon – heat transfer two-phase device for wide application *Arch. Thermodyn.* **36** pp 65–76
[13] Vasiliev L L, Vassiliev L L, Jr, Rabetsky M I, Grakovich L P, Zhuravlyov A S, Shapovalov A V and Rodin A V 2019 Long horizontal vapordynamic thermosyphons for renewable energy sources *Heat Transf. Eng.* **40** pp 258–266

https://doi.org/10.1080/01457632.2018.1426252

[14] Mochizuki M, Nguyen Thang, Mashiko M, Saito Y, Nguyen Tien and Wuttijumnong V 2011 A review of heat pipe application including new opportunities *Frontiers in Heat Pipes* **2** 013001 pp 1–15 https://dx.doi.org/10.5098/fhp.v2.1.3001

[15] Winston Zhang L An Experimental Study on Applying Miniature Loop Heat Pipes for Laptop PC Cooling Technology, Novark Technology Inc. www.novark.com.cn, February 2016

[16] Zhao Y, Yan T and Liang J 2011 Experimental study on a cryogenic loop heat pipe with high heat capacity *Int. J. Heat and Mass Transf.* **54** pp 3304–3308

[17] Hwang G S, Park C W and Kaviany M 2015 High-heat-flux distributed capillary-artery evaporators" in: K Vafai (Ed), *Handb. of Porous Media*, Third ed., Taylor and Francis, CRC Press, pp 597–630

[18] Jinliang Xu, Xianbing Ji, Woiong Yang and Zwei 2014 Modulated porous wick evaporator for loop heat pipes: Experiment *Int. J. Heat Mass Transf.* **72** pp 163–176

[19] Vasiliev L, Zhuravlyov A, Shapovalov A and Litvinenko V 2004 Vaporization heat transfer in porous wicks of evaporators *Arch. of Thermodyn.* **25** 3 pp 47–59

[20] Vasiliev L, Zhuravlyov A and Shapovalov A 2012 Heat transfer enhancement in mini channels with micro/nano particles deposited on a heat-loaded wall *J. Enhanc. Heat Transf.* **19** 1 pp 13–24

[21] Zhi Hu, Yaning Zhang, Bingxi Li, Chi-Chuan Wang and Yongji Li 2018 The influence of the inclination angle and an evaporator wettability on the heat performance of the thermosyphon by simulation and experiment *Int. J. Heat and Mass Transf.* **116** pp 675–684

[22] Franco A and Filippeschi S 2013 Experimental analysis of closed loop two phase thermosyphon (CLTPT) for energy systems *Exp. Therm. Fluid Sci.* **51** pp 302–311

[23] Lahey R T, Jr and Moody F J 1996 *The Thermal-Hydraulic of Boiling Water Nuclear Reactor*, 2nd ed (Illinois: Am. Nucl. Soc.) 631 p

[24] Sviridenko I I and Shevielov D V 2011 Autonomous thermosyphon system for WWER cooldown *Heat Pipe Sci. Technol. An Int. J.* 2 pp 145–159. https://DOI: 10.1615/HeatPipeScieTech.v2.i1-4.150

[25] Esen M and Esen H 2005 Experimental investigation of a two-phase closed thermosyphon solar water heater *Sol. Energy* **79** pp 459–468

[26] Li J, Lin F and Niu G 2014 An insert-type two-phase closed loop thermosyphon for split-type solar water heaters *Appl. Therm. Eng.* **70** pp 441–450

[27] Moradgholi M, Nowee S M. and Abrishamchi I 2014 Application of heat pipe in an experimental investigation on a novel photovoltaic/thermal (PV/T) system *Sol. Energy* **107** pp 82–88

[28] Han L, Shi W, Wang B, Zhang P and Li X 2013 Development of an integrated air conditioner with thermosyphon and the application in mobile phone base station *Int. J. Refrig.* **36** pp 58–69

[29] Perpiña X, Pitón M, Mermet-Guyennet M, Jordà X and Millán J 2007 Local thermal cycles determination in thermosyphon-cooled traction IGBT modules reproducing mission profiles *Microelectron. Reliab.* **47** pp 1701–1706

[30] Sarno C, Tantolin C, Hodot R, Maydanik Yu and Vershinin S 2013 Loop thermosyphon thermal management of the avionics of an in-flight entertainment system *Appl. Therm. Eng.* **51** pp 764–769.

[31] Khrustalev D, Cologer P 2010 Transient modeling of hybrid loop heat pipe systems with multiple evaporators *Front. Heat Pipes.* 1 013002 1–1. DOI: 10.5098/fhp.v1.1.3002

[32] Holman J P 2011 *Experimental Methods for Engineers,* 8th ed (NY: McGraw-Hill) 739 p

Surfactant effect in the thermal performance of a two-phase thermosyphon using Al_2O_3 nanofluid

Karen Cacua[1,2*], Robison Buitrago-Sierra[1] , Bernardo Herrera[1,2], Anderson Gallego[1] and Elizabeth Pabón[2]

[1] Faculty of Engineering, Instituto Tecnológico Metropolitano, Calle 54 A No. 30-01, Medellín, Colombia
[2] Faculty of Sciences, Universidad Nacional de Colombia, Calle 59A 63-20, Medellín, Colombia
[2] Faculty of Mines, Universidad Nacional de Colombia, Calle 80 No. 65-223, Medellín, Colombia

Abstract. An experimental study was carried out to evaluate the effect on the thermal performance of a two-phase closed thermosyphon (TPCT) during operation cycles using nanofluids. Al_2O_3 (0.1 wt.%) based nanofluids with two concentrations of SDBS (0.032 wt.% and 0.064 wt.%) were prepared via two-step method. Stability after nanofluids preparation was evaluated using visual inspection, UV-spectroscopy and dynamic light scattering (DLS). Nanofluids with and without surfactant were used as working fluid in a glass TPCT. Results show that the TPCT thermal resistance was decreased up to 20% respect to water by the presence of nanoparticles and surfactant. A dried limit with water as working fluid was observed and fluid expansion was identified with surfactant presence. Differences in thermal efficiency and temperature profiles using nanofluids with and without surfactant and only surfactant-water as working fluids in TPCT were not significative. However, in evaporator zone, a slight increase of the thermal resistance was identified when surfactant was added. After TPCT operation with nanofluids, nanoparticles agglomeration and sedimentation were observed and a layer on the evaporator surface was observed.

1. Introduction

Heat pipes are effective passive heat transfer devices and are used when heat transfer with small temperature differences is the primary objective. These devices are commonly used in isothermal applications, temperature control, and in heat transfer recovery from low-grade energy waste [1]. Two-Phase Closed Thermosyphons (TPCT) are a simple type of gravity-assisted wickless heat pipes and have applications in a broad temperature range compared to a conventional heat pipe. In the last years, research works on thermosyphons have intensified noticeably due to their abundant applications in heating and cooling processes and solar energy recovering [2].

Nanofluids are a complex class of working fluid that have been used in closed two-phase thermosyphons with the goal of increase their thermal performance. Nanofluids are colloidal

* corresponding author: karencacua@itm.edu.co

IOP Conf. Series: Materials Science and Engineering 1139 (2021) 012012 doi:10.1088/1757-899X/1139/1/012012

suspensions composed by nanomaterials (1-100 nm) dispersed in a fluid. The thermal performance of a thermosyphon using nanofluids has been evaluated experimentally by several researchers. The results found in these investigations shows increases [3–8] or decreases [9–11] in the thermal resistance using different nanofluids. These changes in the thermal resistance of the thermosyphons have been associated with a porous layer deposited on the evaporator surfaces and the effect of this layer in the boiling heat transfer mechanism. This layer is formed due to the agglomeration and deposition of nanoparticles in the heated surfaces on the evaporator. This agglomeration and deposition may perhaps be connected to parameters used when the nanofluid is prepared and complications with the permanence of the nanomaterials disperse with time.

The structure and morphology of this layer is affected by nanoparticle type [5,11], nanoparticle concentration [12] and the chemical additives such us surfactants, used during the nanofluid preparation [13]. Surface active agents or surfactants are commonly used with the goal to improve the nanoparticle dispersion. Parametthanuwat et al. [14] described the effect of oleic acid on the nanofluids performance in a TPCT. Using different concentrations of oleic acid to disperse Ag nanoparticles in water, they founded that with a concentration of 1 w/v% of the surface-active agent, the maximum increase in the TPCT performance was attained. However, the effect of this surface-active agent on the dispersion stability and the mechanisms of heat transfer were not described.

Amiri et al. [4], evaluated experimentally the performance of a TPCT using two types of nanomaterials dispersed in water, functionalized graphene nanoplatelets and graphene nanoplatelets (GNP) with SDBS used as a surfactant. They obtained an increase of 33 % in the heat transfer coefficient using functionalized graphene nanoplatelets compared to the nanofluid using 0.1 wt.% of SDBS surfactant.

The surfactant present in nanofluids could generate modifications in the TPCT thermal behavior [15]. These modifications are associated with variations in surface tension, viscosity and thermal conductivity, which can produce modifications in convection and boiling [16]. However, although is evident that the surfactants can modify the TPCT thermal performance, few studies were found in the literature that evaluated the surfactant presence in the nanofluid and their impact on the TPCT thermal performance.

In this work, an experimental evaluation was carried out to evaluate the effect of surfactant addition to improve the nanoparticles dispersion during nanofluids preparation, on the thermal resistance of a TPCT.

2. Methodology

2.1. Materials and methods

Nanoparticles of Al_2O_3 (<50 nm, from Sigma-Aldrich) were dispersed in deionized water (DI) (18 MΩ•cm) using an ultrasonic processor (Ultrasonicator, QSONICA 500). The mass concentrations of the nanomaterials in water was 0.1% wt. Sodium dodecylbenzene sulfonate (SDBS) was used as surface-active agent at 0.032 and 0.064 wt%. Nanomaterial was dispersed in water by two-step method, first SDBS was added to DI water and mixed using agitation. After, this solution was added to nanoparticles previously weighed. Nanoparticles were dispersed using an ultrasonication probe tip immersed directly in the mix by 30 minutes. The detailed methodology followed to nanofluids preparation is described in a previous work [17].

Nanofluids stability variation with the time was supervised using visual inspection, UV visible spectroscopy (Spectrofotometer, Agilent 8453) and dynamic light scattering technique DLS (Nanoplus HD3, Micromeritics). After evaluating stability in static mode, nanofluids with the best stability were used in a glass TPCT with a coper surface in the evaporator at 40 W and a filling ratio of 45%.

The TPCT experimental setup is shown in Figure 1. A borosilicate glass thermosyphon was used to observe the fluid behavior through operation and the nanomaterial settled after TPCT operation. The TPCT has a length of 400 nm and, an outer and inner diameter of 26 mm and 30 mm, respectively. In the evaporator section, a thermal resistance enfolded along the evaporator was used to bring the power to the working fluid. In the lowest part of the evaporator zone, a copper element was located, with the purpose of examine the Al_2O_3 nanoparticles sedimented after the experiment. Cooling water at 9 °C was used in the condenser section, using a bath to control the temperature (Alpha RA 24, LAUDA). A Coriolis flow meter was used to measure the mass flow of cooling water (Siemens, SITRANS 2100 DI). In the start of the experiments, a vacuum pressure of 12.5 in Hg was used.

Each experiment was done by eight hours and the evaluation was carried out for two days without change the working fluid. To measure the temperature along the TPCT, ten K-type thermocouples were used, two along the evaporator zone, three in adiabatic zone and five in condenser zone as is shown in figure 1. Two more were positioned to evaluate the external water temperature in the condenser. After operation, the evaporator surface was inspected, and nanoparticles deposition was evaluated using a Field Emission Scanning Electron Microscopy (FE-SEM, JEOL JSM7100F).
Thermal resistance was calculated using Equation 1.

$$R_{total} = \frac{\overline{T_e} - \overline{T_c}}{VI} \tag{1}$$

Where, $\overline{T_e}$ and $\overline{T_c}$ are averaged temperatures in the evaporation and condenser zone, respectively. $\dot{m}$ and ΔT are mass flow and temperature difference of the water in the condenser section, respectively; V and I are the voltage and current values of the thermal resistance used to supply the heat in the evaporator section.

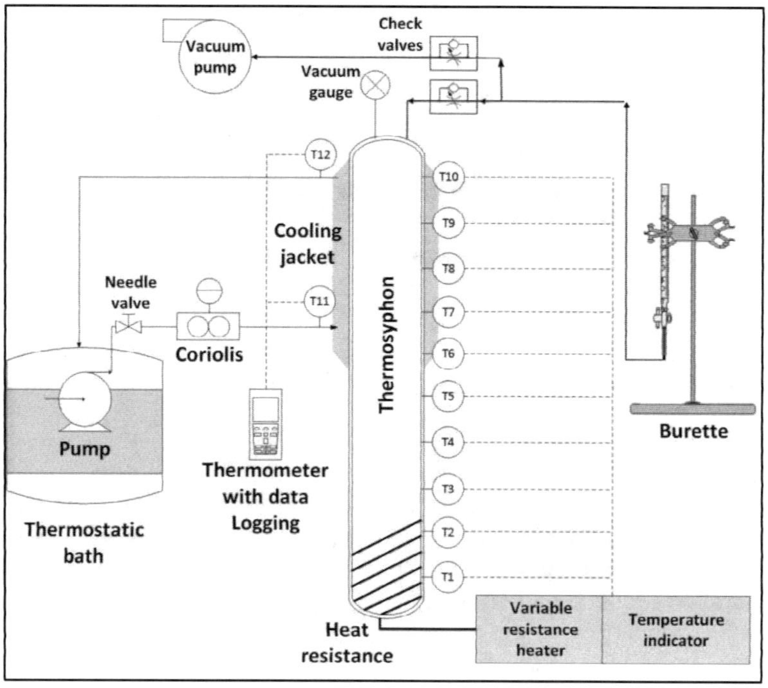

Figure 1. TPCT experimental scheme

3. Results and discussion

3.1. Stability in static mode

Figure 2a, shows the absorbance variation over time for the different nanofluids evaluated. In all cases, absorbance decreases with time, and sedimentation was evident as can be observed in images obtained by visual inspection (see Figure 2b).

According to absorbance variation, the nanofluid most stable was Al_2O_3 0.1 wt.% with SDBS 0.064 wt.%. DLS results showed a hydrodynamic average diameter after 24 days of 121 ±1.25 nm and 117.4 ±1.73 nm for Al_2O_3 0.1 wt.% and Al_2O_3 0.1 wt.% (SDBS 0.064 wt.%), respectively. Both nanofluids were used in the TPCT experimental tests. Additionally, to see the effect of surfactant presence a fluid composed of DI-Water and SDBS was used too.

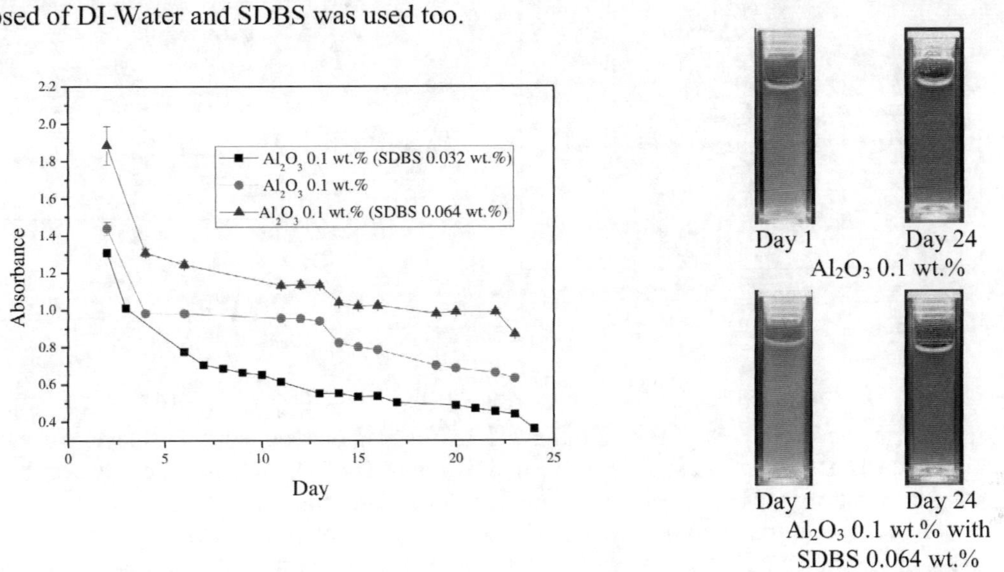

Figure 2. (a) UV-absorbance variation with time **(b)** Visual inspection of nanofluids

3.2. Thermal performance of TPCT

Figure 3, show the thermal resistance (3a) and temperature profiles (3b) of the TPCT, respectively. Results showed decreases up to 20% in the TPCT thermal resistance using Al_2O_3 nanofluids and SDBS-water with respect to deionized water. Water high thermal resistance was caused by a dried zone in evaporator accompanied by an increase in evaporator temperature (see position two in Figure 3b).

Differences in thermal resistance and temperature profiles using nanofluids with and without surfactant and only surfactant-water as working fluids in TPCT were not significative for these experimental conditions. Only a slight increase in evaporator temperatures was observed with the surfactant presence, this can be due to bubbles formation with the decrease in the fluid surface tension. Additionally, when the nanofluid was left in the TPCT without working one day, no changes were identified in the temperature profiles and thermal resistance.

Different phenomena associated with the fluids used in the TPCT can be identified in Figure 4. During the visualization in the operation of the TPCT, an expansion of the fluid and a considerable amount of bubbles were identified with the surface-active agent presence (Figure 5c), this expansion although not affected the thermal performance need to be controlled depending of the experimental parameters used, such as filling ratio, vacuum pressure among others, to avoid problems with the condensed return to the evaporator zone. The effect on heat transfer adding only nanoparticles without surfactant and adding only surfactant to DI water was the same, also, when nanoparticles and surfactant were used simultaneously the effect was not additive and the result was not modified. Additional studies are required to decouple the individual effects.

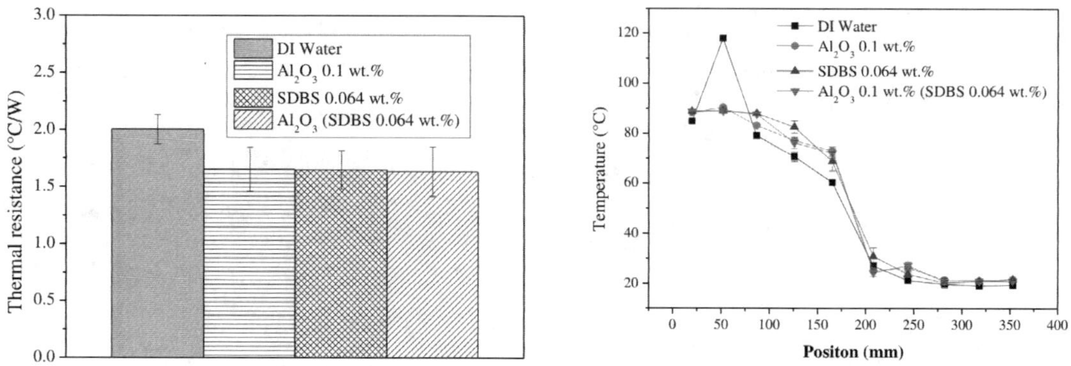

(a) (b)

Figure 3. TPCT (a) thermal resistance and (b) temperature profiles

Effects on heat transfer in TPCT generally have been focused on a nanoparticles layer formation on the evaporator surface, as a result of nanoparticle deposition [8,18–21]. This nanoparticle sedimentation modifies the nucleation sites, and the fluid bubbles generation and surface wettability are altered, affecting the boiling heat transfer phenomena. Moreover, the use of surfactants also contributes to the boiling heat transfer on evaporator zone and this is due to surface tension effect that allow an increase in the nucleation locations and the bubble leaving diameter improving the boiling heat transfer [22–24].

a. DI water, dried zone identified. b. Al_2O_3 0.1 wt.% without dried zone. c. SDBS presence effect.

Figure 4. Different phenomena observed during TPCT operation with different working fluids

After experiments with nanofluids a high sedimentation of nanoparticles was observed, and a layer of Al_2O_3 nanoparticles was identified on the evaporator surface, even with the presence of surfactant (see evaporator surface and SEM image in Figure 5). However, the evident presence of nanoparticles on the evaporator surface and the changes in the boiling phenomena observed in Figure 4 with and without surfactant, do not caused significative changes in the temperature profiles, compared with only the surfactant addition to DI water. However, more studies are necessary to concluded about the effect of surfactant addition to working fluids used in TPCT and the real advantages of nanoparticles dispersion in the common working fluids.

a. **b.**

Figure 5. a. Nanoparticles deposition in evaporator surface. b. SEM image of evaporator surface after operation with alumina nanofluids.

4. Conclusions

Alumina nanofluids with SDBS addition were prepared and their stability was monitored using UV-vis absorbance, DLS and visual inspection. Al_2O_3 with SDBS (0.064 wt.%) showed the best stability in static mode.

The TPCT thermal resistance was decreased up to 20% using Al_2O_3 nanofluids with SDBS and SDBS-water with respect to deionized water.

Differences in thermal resistance and temperature profiles using nanofluids with and without surfactant and only surfactant-water as working fluids in TPCT were not significative for these experimental conditions

Different phenomena with the diverse fluids in the TPCT were observed. A dried zone with DI-water was identified and this was the cause for the poor thermal performance. When surfactant-water was used as a working fluid, a fluid expansion and a big quantity of bubbles were observed, this produced a slight increase in the temperatures in the evaporator.

After experiments with nanofluids a high sedimentation of nanoparticles was observed, and a layer of Al_2O_3 nanoparticle was identified on the evaporator surface, even with the presence of surfactant.

The surface-active agents have an effect in the heat transfer boiling phenomena and in the TPCTs thermal behavior, their effect in TPCT thermal performance must be decoupled from the effect of the nanoparticles. Surfactant represents an alternative to improve the TPCT performance.

5. References

[1] Faghri A 2014 Heat Pipes: Review, Opportunities and Challenges *Front. Heat Pipes* **5**

[2] Jafari D, Franco A, Filippeschi S and Di Marco P 2016 Two-phase closed thermosyphons: A review of studies and solar applications *Renew. Sustain. Energy Rev.* **53** 575–93

[3] Sarafraz M M, Hormozi F and Peyghambarzadeh S M 2014 Thermal performance and efficiency of a thermosyphon heat pipe working with a biologically ecofriendly nanofluid *Int. Commun. Heat Mass Transf.* **57** 297–303

[4] Amiri A, Sadri R, Shanbedi M, Ahmadi G, Chew B T, Kazi S N and Dahari M 2015 Performance dependence of thermosyphon on the functionalization approaches: An experimental study on thermo-physical properties of graphene nanoplatelet-based water nanofluids *Energy Convers. Manag.* **92** 322–30

[5] Kamyar a., Ong K S and Saidur R 2013 Effects of nanofluids on heat transfer characteristics of a two-phase closed thermosyphon *Int. J. Heat Mass Transf.* **65** 610–8

[6] Buschmann M H and Franzke U 2014 Improvement of thermosyphon performance by

employing nanofluid *Int. J. Refrig.* **40** 416–28
[7] Hajian R, Layeghi M and Abbaspour Sani K 2012 Experimental study of nanofluid effects on the thermal performance with response time of heat pipe *Energy Convers. Manag.* **56** 63–8
[8] Shanbedi M, Heris S Z, Amiri A and Baniadam M 2014 Improvement in Heat Transfer of a Two-Phased Closed Thermosyphon Using Silver-Decorated MWCNT/Water *J. Dispers. Sci. Technol.* **35** 1086–96
[9] Chen Y, Wang P and Liu Z 2013 Application of water-based SiO2 functionalized nanofluid in a loop thermosyphon *Int. J. Heat Mass Transf.* **56** 59–68
[10] Xue H S, Fan J R, Hu Y C, Hong R H and Cen K F 2006 The interface effect of carbon nanotube suspension on the thermal performance of a two-phase closed thermosyphon *J. Appl. Phys.* **100** 104909
[11] Khandekar S, Joshi Y M and Mehta B 2008 Thermal performance of closed two-phase thermosyphon using nanofluids *Int. J. Therm. Sci.* **47** 659–67
[12] Yang X-F and Liu Z-H 2012 Flow boiling heat transfer in the evaporator of a loop thermosyphon operating with CuO based aqueous nanofluid *Int. J. Heat Mass Transf.* **55** 7375–84
[13] Grab T, Gross U, Franzke U and Buschmann M H 2014 Operation performance of thermosyphons employing titania and gold nanofluids *Int. J. Therm. Sci.* **86** 352–64
[14] Parametthanuwat T, Rittidech S, Pattiya A, Ding Y and Witharana S 2011 Application of silver nanofluid containing oleic acid surfactant in a thermosyphon economizer *Nanoscale Res. Lett.* **6** 315
[15] Wen D, Lin G, Vafaei S and Zhang K 2009 Review of nanofluids for heat transfer applications *Particuology* **7** 141–50
[16] Hetsroni G, Zakin J ., Lin Z, Mosyak a, Pancallo E . and Rozenblit R 2001 The effect of surfactants on bubble growth, wall thermal patterns and heat transfer in pool boiling *Int. J. Heat Mass Transf.* **44** 485–97
[17] Cacua K, Buitrago-Sierra R, Herrera B, Chejne F and Pabón E 2017 Influence of different parameters and their coupled effects on the stability of alumina nanofluids by a fractional factorial design approach *Adv. Powder Technol.* **28** 37–44
[18] Karimzadehkhouei M, Shojaeian M, Şendur K, Mengï¿½ï¿½ M P and Koşar A 2017 The effect of nanoparticle type and nanoparticle mass fraction on heat transfer enhancement in pool boiling *Int. J. Heat Mass Transf.* **109** 157–66
[19] Mohideen S T and Suresh Kumar R 2016 An Experimental Investigation of the Thermal Performance of Two-Phase Closed Thermosyphon Using Zirconia Nanofluid *Therm. Sci.* **20** 1565–74
[20] Asirvatham L G, Wongwises S and Babu J 2016 Heat Transfer Performance of a Glass Thermosyphon Using Graphene – Acetone Nanofluid **137** 1–9
[21] Shanbedi M, Zeinali Heris S, Baniadam M and Amiri a. 2013 The Effect of Multi-Walled Carbon Nanotube/Water Nanofluid on Thermal Performance of a Two-Phase Closed Thermosyphon *Exp. Heat Transf.* **26** 26–40
[22] Shoghl S N and Bahrami M 2013 Experimental investigation on pool boiling heat transfer of ZnO, and CuO water-based nanofluids and effect of surfactant on heat transfer coefficient *Int. Commun. Heat Mass Transf.* **45** 122–9
[23] Liang G and Mudawar I 2018 Review of pool boiling enhancement with additives and nanofluids *Int. J. Heat Mass Transf.* **124** 423–53
[24] Zhang J and Manglik R M 2005 Additive Adsorption and Interfacial Characteristics of Nucleate Pool Boiling in Aqueous Surfactant Solutions *J. Heat Transfer* **127** 684

Acknowledgement

The authors gratefully acknowledge the financial support of COLCIENCIAS- Colombia (1118-669-46092 CTO: 126-2015), Universidad Nacional de Colombia, and Instituto Tecnológico Metropolitano de Medellín-Colombia

IOP Conf. Series: Materials Science and Engineering 1139 (2021) 012014 doi:10.1088/1757-899X/1139/1/012014

Performance of a small diameter two-phase closed thermosyphon in geyser boiling condition

Nelson Yurako Londoño Pabón[1*], Gabriel Serafin Couto Vieira[1], Kênia Warmling Milanez[1] and Marcia Barbosa Henriques Mantelli[1]

[1] Department of Mechanical Engineering, Federal University of Santa Catarina, Florianopolis 88040-900, Brazil

Abstract. In the present article, the geyser boiling phenomenon is experimentally investigated in a small diameter two-phase closed thermosyphon. The studied thermosyphon has an inner diameter of 4.8 mm and total length of 918 mm. Three specific filling ratios and four temperatures for the thermal bath used to provide cooling water to the condenser were tested. Power was inputted in steps of 20 W, up to 100 W. Variation of the temperatures on the walls of the evaporator, adiabatic and condenser sections as well as variations on pressure, obtained from a pressure transducer installed on the thermosyphon, indicated the presence of Geyser boiling. The results show that the studied variables affect directly the intensity and periodicity of the geyser boiling phenomenon.

1. Introduction

Two-phase closed thermosyphons are high conductance thermal devices. Basically, a thermosyphon can be considered as a closed evacuated container, where a certain amount of working fluid is inserted. It has three sections: evaporator, condenser and adiabatic regions. Heat is inputted in the evaporator and rejected in the condenser section. The working fluid evaporates in the evaporator and the resulting vapor is transported, due to pressure differences, to the condenser region, where heat is then rejected, causing the vapor condensation. The condensate returns to the evaporator region due gravity action, closing the cycle. This heat transfer process is cyclic and passive as it does not require the use of pumps or other active devices, being driven only by temperature differences between evaporator and condenser, which can be quite small. It is imperative that the evaporator section remains under the condenser, as the gravity must be responsible for returning the condensate from the condenser to the evaporator. As stated by Faghri [1], the high thermal conductance of the thermosyphons is directly associated with the high convection coefficients of the evaporation and condensation observed internally the device. These characteristics make thermosyphons attractive solutions to improve the efficiency of heat exchangers, for instance.

However, the same physical phenomena which allow the thermosyphon to work, also limit its maximum heat transfer capacity. These operational limits are important parameters to consider when designing such devices. The literature reports several models and correlations developed to predict the maximum admissible heat power according to the physical phenomena that limits the heat transfer. Many of these correlations are shown in Table 1. Some of these heat transfer limits are resulting from the interaction between the counter flows of the liquid and vapor phases, inside the thermosyphons. By analyzing the equations presented in Table 1, it is observed that these limits are functions of the thermophysical properties of the liquid-vapor phases and of the geometrical properties of the tube. From Table 1 equations, it is also detected that the inner diameter (d_i) is a highly influential parameter.

However, during the tests, it was observed that the geyser boiling phenomenon (from now on referred as GBP) occurred before any other operational limits were reached. Even if it does not represents a limitation to the heat transfer capacity of the thermosyphon, the GBP must be avoided, as it may cause strong vibration and/or may damage the end cap of the condenser if maintained for large periods of time, as stated by Faghri [1].

* corresponding author: nelson@labtucal.ufsc.br

Table 1. Correlations to estimate the thermosyphon operation limits.

Heat transfer limit	Correlation	Authors
Entrainment	$q_{max} = C_w^{\,2}\left[\frac{1}{4}\pi d_i^{\,2.5}\right] h_{lv}[\rho_v g(\rho_l - \rho_v)]^{1/2} / \left[1 + \left(\rho_v/\rho_l\right)^{1/4}\right]^2$ *where* $0.7 \le C_w \le 1.0$	Sakhuja [2]
Entrainment	$q_{max} = C^2\left[\frac{1}{4}\pi d_i^{\,2}\right] h_{lv}\left(\rho_l^{-1/4} + \rho_v^{-1/4}\right)^{-2}[\sigma g(\rho_l - \rho_v)]^{1/4}$ *where* $C = (3.2)^{1/2}\tanh\left(\frac{1}{2}Bo^{1/4}\right)$ *and* $Bo = d_i\left[\frac{g(\rho_l - \rho_v)}{\sigma}\right]^{1/2}$	Tien and Chung [3]
Sonic	$q_{max} = 0.474 h_{lv} A_v (\rho_v p_v)^{1/2}$	Busse [4]
Viscous	$q_{max} = r_v^{\,2} h_{lv} A_v \frac{\rho_v p_v}{16\mu l_{ef}}$	Busse [4]
Boiling	$q_{max} = 0.12 h_{lv} A_e \rho_v^{\,1/2}[g\sigma(\rho_l - \rho_v)]^{1/4}$	Groll and Rosier [5]

In the present paper, the heat transfer performance of a small diameter thermosyphon is studied, with special attention to the occurrence of the GBP.

2. Literature review

As stated by Faghri [1], the GBP happens when the input power on the evaporator is insufficient to allow constant nucleate boiling. In this case, the temperature of the liquid pool increases until it becomes superheated. A vapor bubble then nucleates, somewhere in the liquid pool and its size quickly increases up to the inner diameter of the tube. This vapor bubble separates a portion of the working fluid from the rest of the pool. The bubble continues to grow until it is inside pressure is enough to impel this portion of fluid upwards, which reaches the top of the condenser. Then, a liquid slug returns to the evaporator in the form of subcooled liquid film over the wall. This subcooled liquid cools down the temperature of the liquid pool. This cycle configures the GBP. Afterwards, a certain time is required until the liquid pool can reach superheated condition again and a new cycle of the GBP can happens. A representation of GBP is shown in Figure 1.

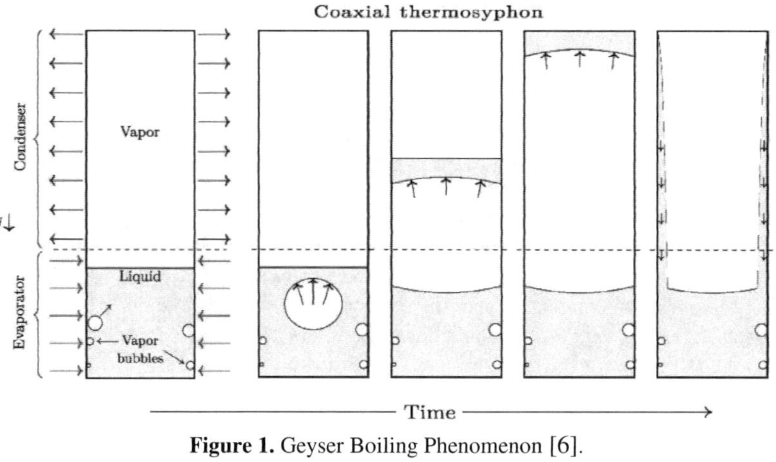

Figure 1. Geyser Boiling Phenomenon [6].

Gross [7], based in 2529 experimental points, studied pool boiling in a thermosyphon. The heat transfer was classified in two categories: two-phase natural convection regime (undeveloped boiling) and nucleated boiling (developed bubble boiling). His results agreed with those of Semena and Kisselev [8], who associated the undeveloped boiling with low input power and the developed bubble boiling with high input power levels.

Andros and Florschuetz [9] observed that film boiling happens within the condensate liquid film, which then breaks in steady streams as it runs down the inner walls of the thermosyphon. The wall high temperature cause nucleation of bubbles as these streams approach the evaporator, inducing instabilities in the streams.

Negishi and Sawada [10] experimentally studied the thermal performance of an inclined two-phase closed thermosyphon. The involucre was composed by a 330 mm long copper tube with an inner diameter of 13 mm. Water and ethanol were selected as the working fluids. The inclination angle varied between 20° and 40° for the water-filled and between 25° and 45° to the ethanol-filled thermosyphons. According to their results, filling ratios (ratio between working fluid volume and evaporator volume) from 25 to 60% for water and from 40 to 75% for ethanol yielded the better heat transfer rates. They also concluded that GBP takes place for filling ratios greater than 70 percent.

Negishi [11] observed that the periodicity of the GBP depends on the condenser section length and on the input power at the evaporator. He observed that longer condenser sections yielded GBP with longer periods. In turn, as the input power in the evaporator increased, the GBP period decreased, and the intensity with which the fluid hit the condenser top was also attenuated. Lin et al. [12] confirmed that smaller filling ratios and shorter evaporator lengths make the periodicity of the GBP shorter, declining almost linearly with the input power increase. The amplitude of the oscillations was also found to decrease as input powers increase.

Mantelli et al. [13] experimentally studied the behavior of a two-phase closed thermosyphon. Water was used as the working fluid. The involucre was made of a 1220 mm long stainless steel tube with an inner diameter of 17.05 mm. The results confirmed the occurrence of GBP for input powers lower than 25 W. On the other hand, for higher input powers, nucleated pool boiling intensified and GBP disappeared. These results agree with Faghri [1], who stated that, even if GBP cannot be qualified as a thermosyphon limit, it is a startup phenomenon, which affects its performance.

Kunkoro et al. [14] studied the geysering mechanism in a two-phase closed thermosyphon. A 2507 mm long, transparent glass tube was used to observe the phenomenon. It was discovered that the temperature distribution or the internal-energy storage pattern might play an important part on the GBP onset. The temperature distribution depends on the geometry of the system, on the physical properties of the working fluid and on the heat transfer process. They justified the GBP mainly to the superheating of the fluid. During the nucleated boiling, the GBP could still be observed even with the vapor pressure rising.

Khazaee et al. [15] studied the influence of four parameters on GBP: filling ratio, input power, mass flow rate through the condenser refrigeration system and aspect ratio (ratio between evaporator length and tube inner diameter). They observed that, increasing the filling ratio, the period and the intensity of the oscillations also increased. However, for filling ratios smaller than 30 percent, the GBP disappears. Moreover, with an increase in the input power, the period and the intensity of the oscillations decreased, until they disappeared completely. For smaller aspect ratios and mass flow rates through the condenser refrigeration system, the period of oscillation increases.

Xia et al. [16] observed that the heat transfer instabilities in the thermosyphon evaporator are mainly caused by the behavior of bubbles, fluid physical properties and operating pressures. They classified the heat transfer in the thermosyphon in three modes: natural convection, intermittent boiling (GBP) and fully developed nucleated boiling.

Smith et al. [17] proposed the first multiphase flow map for closed two-phase thermosyphons, identifying four types of flow: Geyser Flow, Slug/Plug Flow, Churn Flow and Bubbly Flow. The map was designed according to the degree of confinement and the rate of steam production, establishing four zones with the four boiling regimes. Intermittent boiling was observed in high levels of confinement and high heat flows, in which relatively large bubbles are characteristics of the Geyser Flow regime. However, in conditions of low confinement at high steam production rates, the boiling regimes are similar to those observed in the pool boiling (Bubbly Flow). Small diameter thermosyphons show boiling regimes different from those commonly reported for the conventional thermosyphons (pool nucleation and film condensation).

3. Experimental procedure

The thermosyphon studied in the present work was made from a 918 mm long copper pipe with an inner diameter of 4.8 mm. Its three regions: evaporator, adiabatic and condenser, had the lengths: 200, 90 and 628 mm, respectively. The working fluid was chosen according to the suggestion of Reay et al. [18], who affirmed that working fluids with a high merit number should be preferred to minimize the gradient of temperature over the thermosyphon. For the operating temperature range of this thermosyphon, water was selected as working fluid because it exhibited the highest merit number. Four cartridge electric resistances were inserted in an aluminum block in close contact with the evaporator, to provide the thermosyphon input power by means of a power supply (MEC 1310). The evaporator and adiabatic sections were thermally insulated from the environment using expanded vermiculite. The heat rejected by the condenser is dissipated to a water jacket. The condenser section and the water jacket form a tube and shell heat exchanger. The water jacket is also connected to a rotameter (Omega FL1504A) and a waterbath (Lauda RUK-40S). These devices are responsible for keeping the cooling fluid mass flow rate and the temperature constant. Figure 2 depicts the diagram of the workbench.

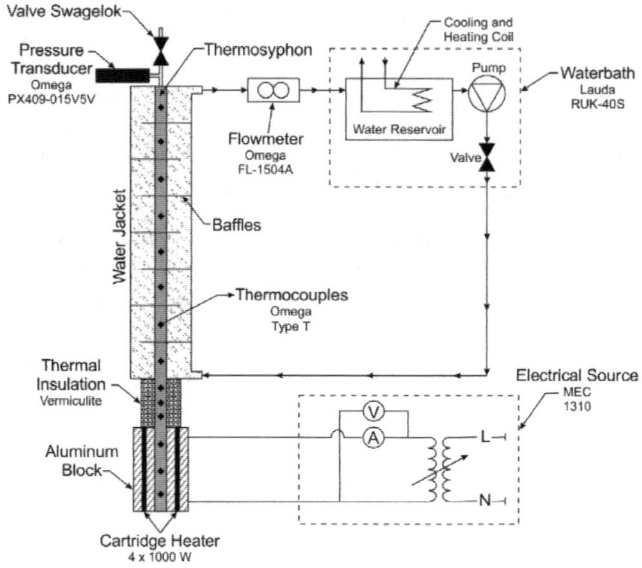

Figure 2. Workbench diagram.

A pressure transducer was installed on the thermosyphon (Omega PX409-015V5V), along with fourteen type T thermocouples, as shown in Figure 3. The workbench also counts with three more thermocouples: one installed on the aluminum block and other two on the entry and exit of the water jacket.

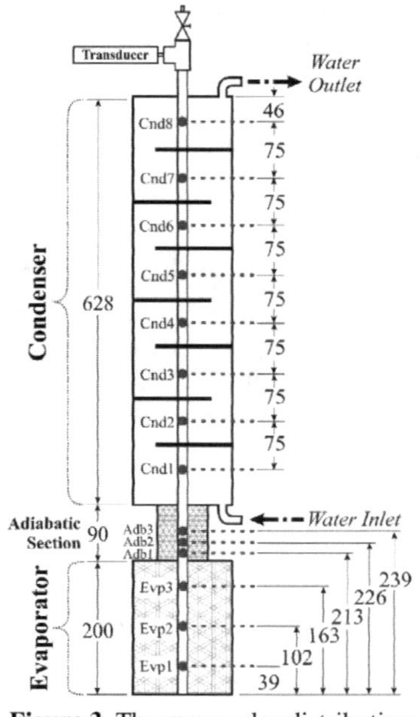

Figure 3. Thermocouples distribution.

IOP Conf. Series: Materials Science and Engineering 1139 (2021) 012014 doi:10.1088/1757-899X/1139/1/012014

All measurement instruments were calibrated aiming to obtain an accurate evaluation of the thermosyphon behavior. The uncertainties related to each measurement device used during the experiments are showed in Table 2. The propagation of these errors for indirect measurements was determined according to equation 1, following the methodology of ISO/IEC GUIDE 98-3:2008 [19].

$$u_c(f) = \sqrt{\sum_{i=1}^{N} \left(\frac{\partial f}{\partial x_i}\right)^2 u^2(x_i)} \quad (1)$$

In this equation, x_i represents each variable, on which the function, f, is dependent. The errors derived from input power (q_w) and thermal resistance calculations using the Eq. (1) were 2.8%.

Table 2. Uncertainty associated with experimental measurements.

Measurement	Device	Experimental uncertainty
Temperature	Omega T-type thermocouple	±1.5 °C
Pressure	Omega pressure transducer	±1,8%
Voltage & Current	MEC 1310 power supply	<0.2%
Rotameter	Omega FL1504A	±2,0%

The curves for correlations, obtained using the expressions presented in Table 1 that predict the operational limits of the studied thermosyphon, are depicted in Figure 4. From this figure, it is possible to observe that the entrainment limit is the first to be reached.

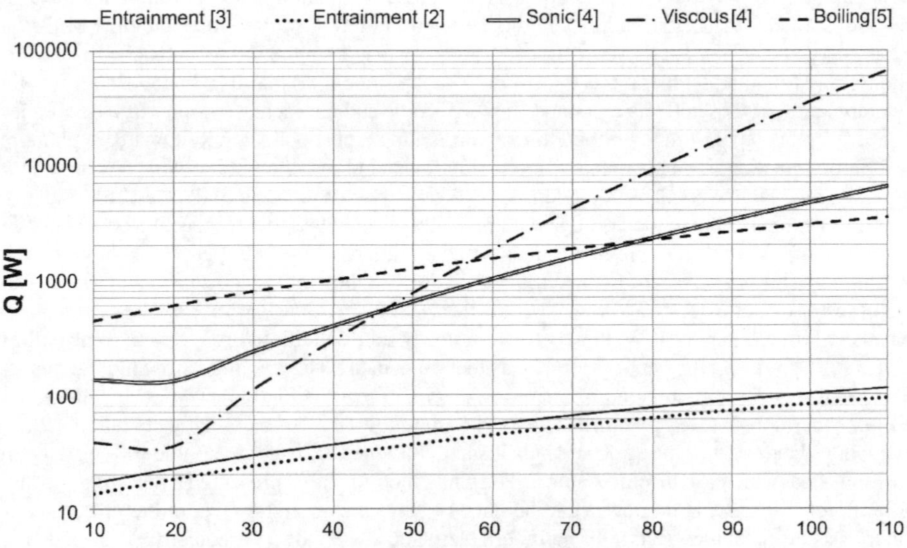

Figure 4. Operation limits *versus* saturation temperature.

For the present experimental work, the input powers ranged from 20 to 100 W with increments of 20 W. Each power step was kept constant until steady state conditions were reached.

Three filling ratios were selected: 35%, 66% and 90%, with which the liquid pool upper level reached the thermocouples Evp1, Evp2 and Evp3, respectively (see Figure 3).

The water, which circulates through the water jacket, was kept at a constant flow rate of 4.62 l/min, and the tests were carried out for two different temperatures: 20 and 30 °C. Table 3 presents an overview of the test conditions.

Table 3. Planning of experiments.

Test	Filling Ratio [%]	Water jacket temperature [°C]
1	35	20
2	35	30
3	66	20
4	66	30
5	90	20
6	90	30

4. Results and discussion

Figure 5 presents the obtained data of temperature and pressure as a function of time, for each one of the six tests. The temperature range (vertical axis) is the same for all plots while small changes can be observed for the time range (horizontal axis). The thermocouples are labeled according to Figure 3. Besides, Twin and Twout are the inlet and outlet water bath temperatures, Tresist is the electrical resistance temperature and P_trans is the working fluid pressure.

In the beginning of test 2, it is perceived that the pressure peaks at near 3000 Pa and then, suddenly decreases. This happens because the temperature of the thermal bath (Twin and Twout) and the temperature of the condenser are higher than the temperatures of the evaporator and adiabatic sections. Under this condition, the thermosyphon cannot work and the heat input merely increases the vapor pressure. When the temperatures of the evaporator and adiabatic sections reach the temperature of thermal bath, the device reaches thermal equilibrium and, at this point, the pressure also lowers until it reaches a equalized level.

This condition of non-operation is also observed in tests 1, 4 and 6. However, for these cases, pressure only grows. For the aforementioned tests, the condition of non-operation is temporary. Therefore, in the present analysis, only the phenomena that happen after the operation conditions are reached (all thermosyphon temperatures are above T_{win} and T_{wout}) will be analyzed.

From Figure 5, it is possible to observe that in tests 3 to 6, the temperature and pressure data present very short period oscillations. In order to better understand this behavior, Figure 6 shows a data window of the 5th test (see Table 3) for the 40 and 100 W input power levels. It is noted that, at around 8500 seconds, the first sudden increase in the condenser pressure and temperatures is observed, at the same time that Evp3 shows a sudden temperature drop. The combination of these effects evidences the GBP. The sudden increase in pressure occurs because, due to its high pressure, the vapor within a bubble bursts, propelling a portion of liquid upwards with a great deal of momentum. Before bursting, the superheated liquid slug located above the bubble which volume is increasing, reaches its higher height in the condenser, evidenced by the continuous increase of the Cnd1 to Cnd5 temperatures with time. Due to the bubble burst, the liquid above the bubble is expelled and reaches the cooled condenser wall, cooling down quickly. After the bubble burst, the subcooled liquid occupies the previous vapor bubble space in the evaporator wall. This is evidenced by the temperature drop on thermocouple Evp3. The GBP is also observed with greater or lesser extent for tests 3 to 6.

It is also gathered from Figure 6, 40W, that, before the steady state conditions are reached, some of the condenser thermocouples still do not feel the vapor presence, as they remain at water cooling bath temperatures (around 20o C). However, as the evaporator temperature increases, the vapor reaches high levels in the condenser, observable by the one by one thermocouple temperature increase, which follows the order of their installation in the condenser, from the bottom to the top. Until the 7500 seconds instant, only the Cnd1 thermocouple followed the trend of the adiabatic section (steady increase in temperature). The Cnd2 thermocouple presented a pulsating reading, showing that the GBP effect reached this thermocouple height. At Cnd3 height and above, the temperatures present the same behavior as the adiabatic section. Thermocouples Cnd4 to Cnd6 sporadically presented the pulsating behavior characteristics.

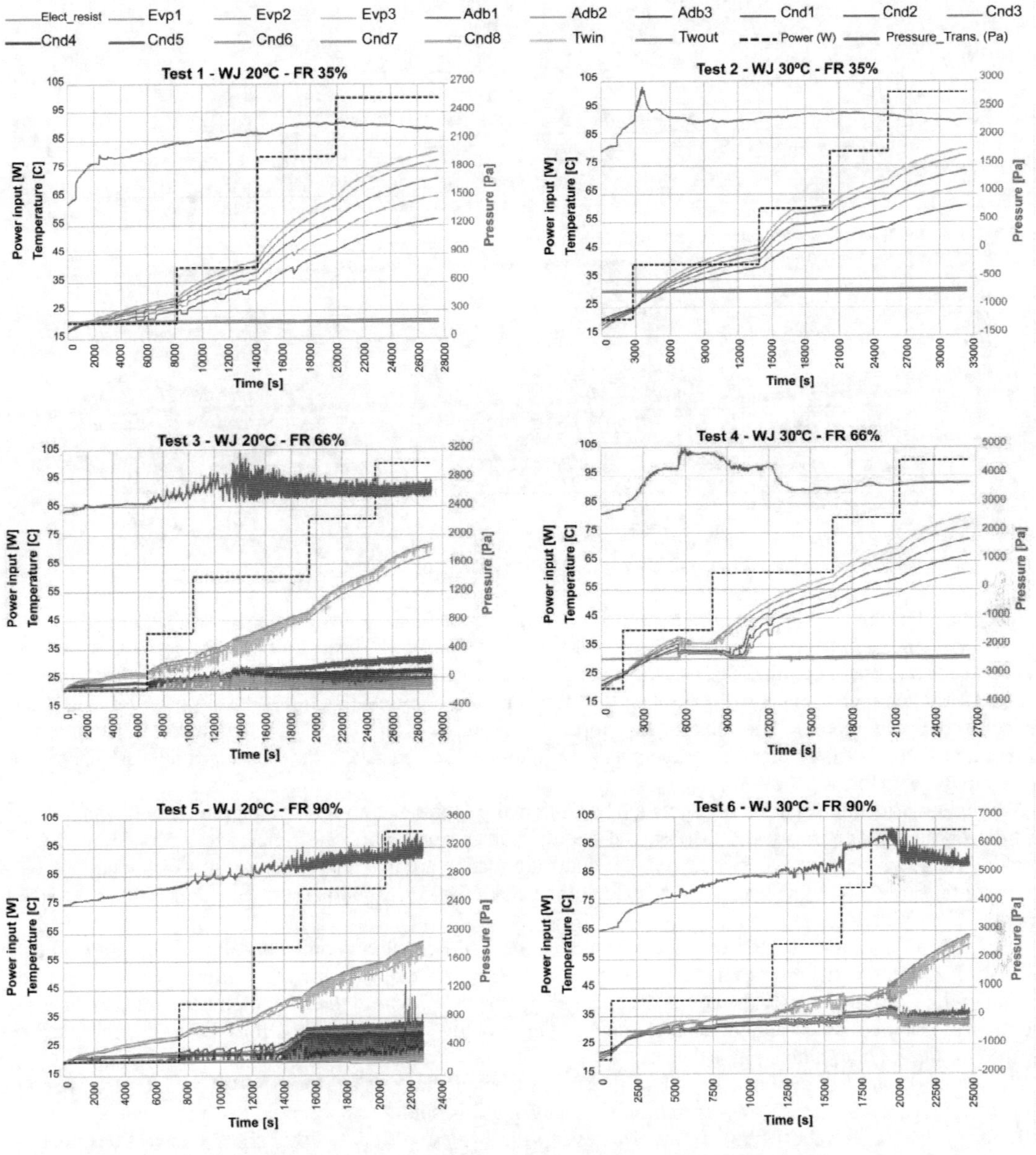

Figure 5. Test results for the small diameter thermosyphon.

Furthermore, when analyzing the temperature readings for the 100 W input power level (Figure 6 right plot), it is possible to observe intense and high frequency oscillations of the temperature readings, for all the thermocouples of the condenser section.

Figure 5 also shows that, despite of the reading oscillations, significant temperature differences between the condenser and thermal bath are observed for the thermosyphon under tests configuration 3, 5 and 6, indicating that the vapor reaches the superior end of the device. This means that the thermosyphon works better under these operation conditions, with lower thermal resistances. The temperature data of test 6 configuration shows that, in this case, the full potential of the condenser section is reached (all the condenser is heated by the vapor). Therefore, the thermal resistance is lower and, as a consequence, the temperatures reached in all the sections are lower.

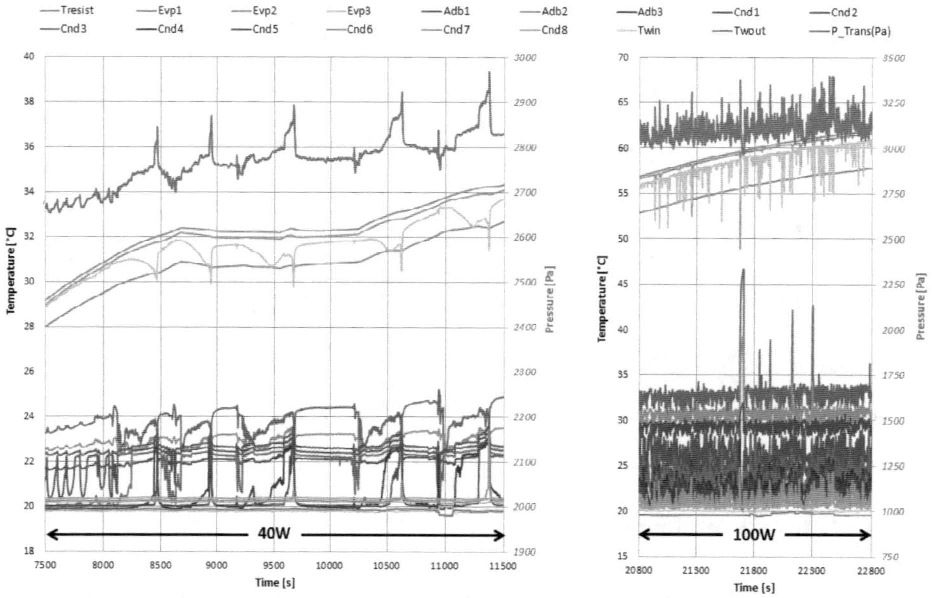

Figure 6. Test 5th for 40 and 100 W of input power.

As depicted in Figure 5, the steady state for the 40W power step of the test 4 presents similar behavior as the one observed in test 6. However, as the input power increased to 60W, the behavior becomes unstable with a drop of condenser's temperature readings, as well as an increase of temperature readings of the adiabatic section and evaporator. After approximately 13000 seconds of test run, it is observed that all of the temperature readings of the condenser section present the value as the temperature readings of the water bath, regardless of input power.

In order to help understanding these phenomena, the thermodynamic state of the working fluid was analyzed, as depicted in Figure 7.

As previously mentioned, test 6 presents the best thermal performance. By analyzing the thermodynamic state of the working fluid for this test, it is observed that the points representing the condenser section show that the working fluid is subcooled, while the points for adiabatic section are found over the saturation line and the ones for evaporator, over the superheated vapor region of the chart. This distribution represents the best thermosyphon operation conditions.

For the thermodynamic state of tests 3 and 5 (see Figure 7), the evaporator and adiabatic section points show a rightward shift in the graphs with increment in power. This causes the adiabatic section to move away from the saturation line. Simultaneously the points of the condenser, located near to the saturation line, are also displaced rightwards, leading them to a more unstable region. This would explain the reason for GBP being characterized by the chaotic and intermittent phase changes.

The thermodynamic state of test 4 explains the steady state condition found for the 40W power step, where the thermosyphon exhibits satisfactory operation. By analyzing this point, it is observed that the condenser and the adiabatic section are on the saturation line. However with increase of input power, these two sections, together with the evaporator, present a thermodynamic state of superheated steam. This shows the blockage of the thermosyphon.

The thermosyphon blockage is also observed in tests 1 and 2. Figure 5 shows that the condenser thermocouples readings presented the same temperature as thermal bath, while the ones pertaining adiabatic and evaporator sections increased their readings, with the increase of the input power. By analyzing this phenomena, in light of the thermodynamic states (see Figure 7), the working fluid in all three sections of the thermosyphon is indicated to be in superheated vapor state. This results in a blockage of the thermosyphon, thus preventing thermal exchange.

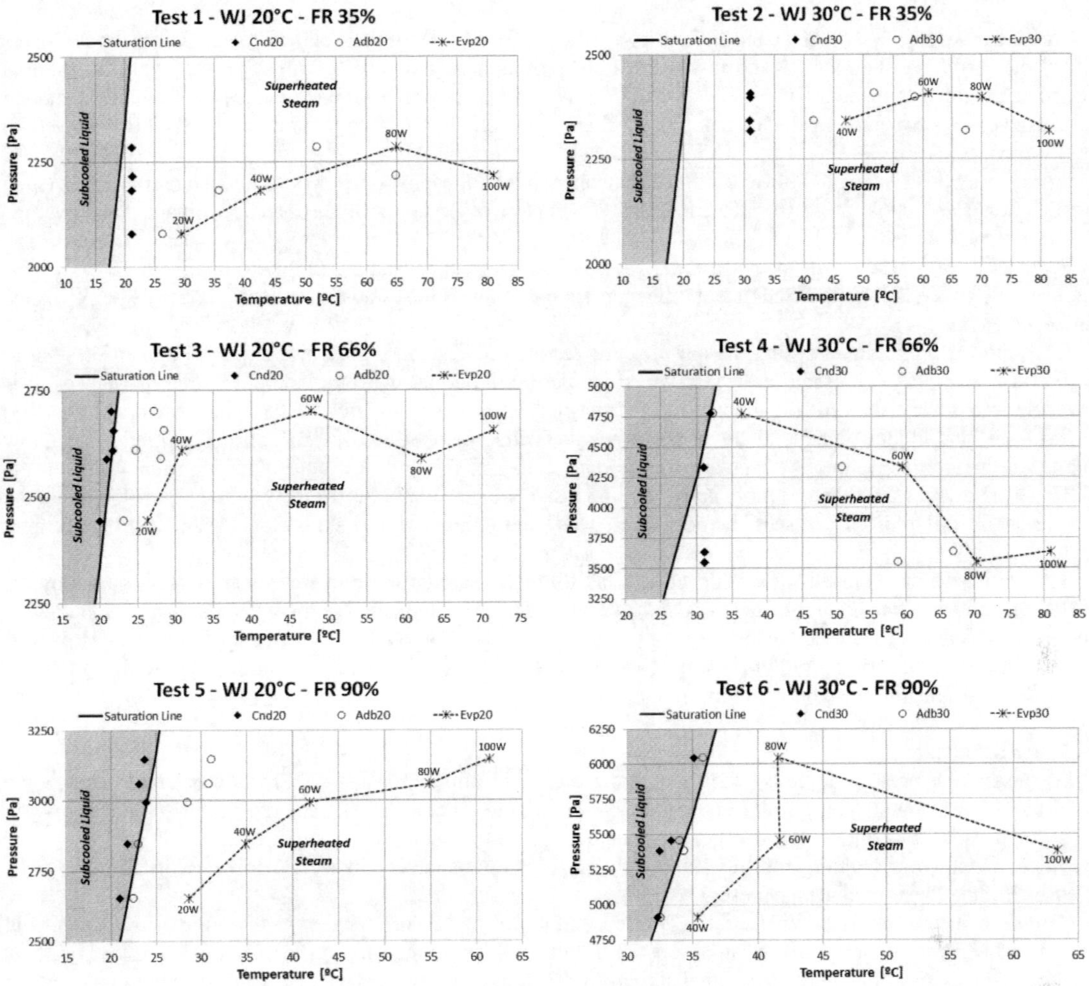

Figure 7. Thermodynamic state of the tests.

Figure 8 presents the thermal resistance as a function of the input power for all of the six tests.

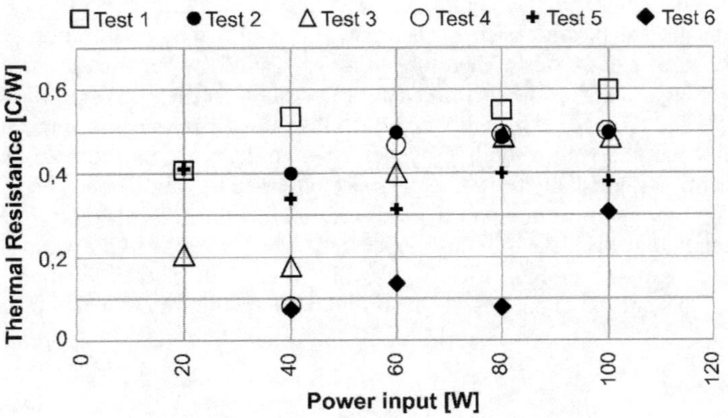

Figure 8. Thermal resistance of the tests.

In Figure 8, it is observed that test 6 presents average values of 0.08 C / W of thermal resistance, for the input powers of 40 and 80W. Furthermore, some instabilities are presented at the 60W of input power, with the thermal resistance being calculated as 0.13 C/W. For the 100W of input power, its value reaches 0.3 C/W. This sudden increased can be explained by the tendency of the thermosyphon to dryout, even when GBP (See Figure 5, test 6) is still present.

From Figure 8, it is also possible to conclude that when the thermosyphon reaches the blockage conditions (such as the ones from tests 1, 2 and from test 4 after 40W input power), the thermal resistance remains at an average value of 0.5 C/W.

For tests 3 and 5, where GBP is evident, the values of the thermal resistances range from 0.2 to 0.45 C/W. This indicates that even though the GBP is not a phenomenon that limits heat transfer, it does hamper its performance and efficiency.

As discussed above, in analyzing the temperature versus time results, Figure 5, it is possible to ensure that in tests 3 to 6 the GBP is present. However, when comparing these data with the global thermal resistances, it is possible to associate that, for higher values of resistance, a greater oscillation of the data of temperatures and pressures occurs. On the other hand, lower power inputs yield longer periods of GBP. This is justified since in low input power rates, more time is necessary for the overheating of the pool of liquid. For higher input power, the GBP period decreases, as there is more active boiling inside the pool. This causes more liquid to be blown into the condenser, resulting in a decrease in the pool level. This leads to higher thermal resistance values for higher input powers.

By comparison of the results, it was observed that, for higher condenser temperatures, the liquid blown by the GBP can reach the total height of the condenser section (Cnd8). This increases the condenser area available for thermal exchange. As a consequence, there is a decrease in the thermal resistances for both filling ratios. However, when comparing both filling ratio operation conditions, it is perceived that increasing the fill rate, from 66% to 90%, decreases the thermosyphon's thermal resistance.

5. Conclusions

After the six tests presented in this paper, with the prescribed filling ratios and condenser temperatures as shown in Table 3, it was found that, for all tests with a 35% filling ratio (tests 1 and 2), the thermosyphon did not work. This happens because the working fluid was found to be in the superheated vapor state in all sections of the thermosyphon. The remaining four tests showed that the thermosyphon did work. However, in all these tests, Geyser Boiling Phenomenon was present.

GBP is a mechanism that disturbs the thermosyphon's capacity of transfer heat. Due to its chaotic and intermittent characteristics, it causes an abrupt change in the internal pressure of the tube and in the temperatures of the three regions of the thermosyphon. This causes the evaporator to overheat and the thermal resistance of the device to rise.

It has been shown that with increasing the water jacket temperature, the vapor front is able to reach higher positions on the condenser. This causes the exchange area to be larger, enhancing the heat transfer.

Contrary to the work of Negishi and Sawada [10] the GBP was present in all filling ratios and not only in filling ratios higher than 70% as they claim. However, it must be taken into account that the diameter tested in this work is 4.8 mm, much smaller diameter than the 13 mm used in their thermosyphon.

The small internal diameter makes it easier for the bubble to occupy the cross-section area of the evaporator, which facilitates the occurrence of the GBP. Therefore, in the present thermosyphon, this phenomenon was present in all of the tests that brought together the thermodynamic conditions for it to operate as such. Unlike Lin et al. [12], Mantelli et al. [13] and Khazaee et al.[15] who affirm that this phenomenon occurs only at low heat flows, the present results show that the Geyser boiling phenomenon is always present due to the confinement effect caused by the small thermosyphon diameter, which was also observed by Smith et al. [17].

Of all the tests performed, only in three power levels did the thermal resistance reached values smaller than 0.1 °C/W. For these conditions, the GBP is almost imperceptible. Although the GBP does not limit the operation of the thermosyphon, it does hinder the efficiency of this equipment. Therefore, more in-depth studies are necessary to understand this phenomenon, how it occurs and what are the factors which influence its generation.

6. Nomenclature

q_w : Heat rate (W)
C_w : Experimental constant
d_i : Internal diameter (m)
h_{lv} : Enthalpy of vaporization (J/kg)
ρ_v : Specific mass of the vapor (kg/m³)
ρ_l : Specific mass of the liquid (Kg/m³)
g : Gravitational constant (m/s²)
σ : Surface tension (N/m)
Bo : Bond number
A_v : Cross-sectional steam area (m²)
p_v : Steam pressure (Pa)
r_v : Radius of the steam area (m)
μ : Dynamic viscosity (Pa.s)
l_{ef} : Effective length (m)
A_e : Radial evaporator area (m²)
u_c : Combined uncertainty

REFERENCES

[1] Faghri A 1995 *Heat Pipe Science and Technology* (1st ed. Taylor & Francis, Washington, D.C., Ch. 6)
[2] Sakhuja R 1973 Flooding constraint in wickless heat pipes *ASME Publ.* 73-WA/HT-7
[3] Tien C and Chung K 1979 Entrainment Limits in Heat Pipes *AIAA Journal* **17** 6 643-646
[4] Busse C 1973 Theory of the ultimate heat transfer limit of cylindrical heat pipes, *Journal International Heat Mass Transfer* **16** 169-186
[5] Groll M, Rosier S 1992 Operation Principles and Performance of Heat Pipes and Closed Two-Phase Thermosyphons (Review) *J. Non-Equilib. Thermodyn.* **17** 121
[6] Tecchio C, Oliveira J L G, Paiva K V, Mantelli M B H, Galdolfi R and Ribeiro L G S 2017 Geyser Boiling phenomenon in two-phase closed loop-thermosyphons *International Journal of Heat and Mass Transfer* **117** 29-40
[7] Gross U 1990 *Pool Boiling Heat Transfer inside a Two-Phase Thermosyphon-Correlation of Experimental Data* (9th International Heat Pipe Conferenc Los Alamos (NM)) **2** 57–62
[8] Semena M G and Kiseelev Yu F 1978 Heat-Exchange Processes in the Heat-Supply Zones of Two-Phase Thermosyphons Operating on Freon 11, 113, and 142 and on Water and Ethanol, *Inzhenerno - Fizicheskii Zhurnal (Russian translated)* **35** 2 211-217
[9] Andros F E and Florschuetz L W 1983 The two phase closed thermosyphon: an experimental study with flow visualization *Two Phase Transport and Reactor Safety* **IV** 111–129
[10] Negishi K and Sawada T 1983 Heat transfer performance of an inclined two phase closed thermosyphon, *Int. J. Heat Mass Transfer* **26** 8 1207–1213
[11] Negishi K 1983 *Unstable Phenomena in a Two-Phase Closed Thermosyphon* (Proc. Symposium on Mechanics for Space Flight, Report S.P., Institute of Space and Astronautical Science Tokyo, Japan) **1** 257–263
[12] Lin T F, Lin T, Tsay Y L, Wu J C and Shyu R J 1995 Experimental Investigation of Geyser Boiling in an Annular Two-Phase Closed Thermosyphon *International Journal Heat and Mass Transfer* **38** 295–307
[13] Mantelli M H B, De Carvalho R, Colle S and de Moraes D 1999 *Study of closed two-phase thermosyphons for bakery oven applications* (National Heat Transfer Conference; NHTC99-205; American Society of Mechanical Engineers (ASME))
[14] Kunkoro H, Rao Y F and Fukuda K 1995 An experimental study on the mechanism of geysering in a closed two-phase thermosyphon *Int. J. Multiphase Flow* **21** 6 1243–1252
[15] Khazaee I, Hosseini R and Noie S H 2010 Experimental investigation of effective parameters and correlation of geyser boiling in a two-phase closed thermosyphon *Applied Thermal Engineering* **30** 5 406-412
[16] Xia G, Wang W, Cheng L, Ma D 2017 Visualization study on the instabilities of phase-change heat transfer in a flat two-phase closed thermosyphon *Applied Thermal Engineering* **116** 392-405
[17] Smith K, Kempers R and Robinson A J 2018 Confinement and vapour production rate influences in closed two-phase reflux thermosyphons Part A: Flow regimes, *International Journal Heat Mass*

doi:10.1088/1757-899X/1139/1/012014

Transfer **119** 907-921
[18] Reay D, McGlen R and Kew P 2013 *Heat Pipes - Theory, Design and Applications* (6th ed., Butterworth Heinemann, Elsevier) 52
[19] ISO/IEC GUIDE 98-3:2008. Uncertainty of measurement — Part 3: Guide to the expression of uncertainty in measurement (GUM:1995)

Acknowledgement
The authors would like to thank the members of the Heat Pipe Laboratory of the Federal University of Santa Catarina (LABTUCAL-UFSC) for their collaboration during the progress of the work.

AUTHOR INDEX

Institute of Physics
Dirac House, Temple Back
Bristol BS1 6BE UK

ISSN: 1757-8981
ISBN 978-1-7138-3414-4

9 781713 834144

CURRAN ASSOCIATES INC.
proceedings.com